Governing a Common Sea

Governing a Common Sea

Environmental Policies in the Baltic Sea Region

Edited by
Marko Joas, Detlef Jahn
and Kristine Kern

publishing for a sustainable future

London • New York

First published by Earthscan in the UK and USA in 2008

Typeset by JS Typesetting Ltd, Porthcawl, Mid Glamorgan

Cover design by Yvonne Booth

For a full list of publications please contact:

Earthscan
2 Park Square, Milton Park, Abingdon, Oxfordshire OX14 4RN
Simultaneously published in the USA and Canada by Earthscan
711 Third Avenue, New York, NY 10017

First issued in paperback 2014

Earthscan is an imprint of the Taylor and Francis Group, an informa business

Earthscan publishes in association with the International Institute for Environment and
Development

A catalogue record for this book is available from the British Library

Library of Congress Cataloging-in-Publication Data

Governing a common sea : environmental policies in the Baltic Sea Region / edited by
Marko Joas, Detlef Jahn, and Kristine Kern.
    p. cm.
  ISBN 978-1-84407-537-9 (hardback)
  1. Environmental policy–Baltic Sea Region–International cooperation. 2. Baltic
States–Politics and government–1991– I. Joas, Marko. II. Jahn, Detlef. III. Kern, Kristine.
  GE160.B29G68 2008
  363.7'0526091821–dc22

                       2008001739
ISBN 13: 978-1-84407-537-9 (hbk)
ISBN 13: 978-1-138-00204-3 (pbk)

Contents

List of Figures and Tables

FIGURES

TABLES

List of Contributors

Tuula Aarnio, PhD, was programme manager for the Finnish Baltic Sea Research Programme (BIREME) (2004 to 2007) at the Academy of Finland. Currently, she is a seconded national expert (SNE) at the European Research Council (ERC), Brussels, Belgium, where she works as science officer in the S2 Unit of the Dedicated Implementation Structure.

Yrjö Haila received his PhD in ecological zoology at the University of Helsinki, Finland, in 1983, and has been professor of environmental policy at the University of Tampere, Finland, since 1995. His publications include articles on bird ecology; conservation biology; social and philosophical dimensions of ecology and environmental issues; environmental policy; and humanity and nature.

Ann-Sofie Hermanson is a lecturer in political science at Åbo Akademi University, Finland, where she also earned her doctoral degree. Her research interests are environmental politics and administration according to a comparative perspective, environmental ideologies and political theory.

Detlef Jahn has been professor of comparative politics at the University of Greifswald, Germany, since 1999. He earned his PhD at the European University Institute in Florence, Italy. His research interest lies in comparative environmental policy and modelling domestic political processes and globalization.

Marko Joas earned his doctoral degree in 2001 and is currently working as the head of research in public administration at Åbo Akademi University, Finland. His research interests include developing forms of democracy and governance, especially in a comparative environmental policy and administration setting.

Kristine Kern is currently a visiting professor at Södertörn University College, Sweden. She holds a PhD in political science from the Free University of Berlin, Germany. Her main research areas are environmental policy and sustainable development, governance in multilevel systems, European Union studies and urban studies.

Aija Kettunen is research director at the University of Kuopio, Department of Health Policy and Management, Finland. Her doctoral thesis was on local environmental conflicts, presented in 1998 at Åbo Akademi University. She is currently working on social economics and social policy issues.

Panu Kontio works in the Finnish Environment Institute as a senior researcher. Since 1993, he has been the manager of several bilateral and multilateral projects in a number of Baltic states. He received his MSc in land resource economics and environmental management at the University of Helsinki, 1989, and is currently researching community structure issues.

Kati Kuitto is research fellow at the Department of Political Science at the University of Greifswald, Germany. Her research interests include transitional policy developments in Central and Eastern Europe, comparative welfare state research and the methodology of comparative political science.

Simo Laakkonen is a senior researcher at the Department of Social Science History, University of Helsinki, where he also earned his doctoral degree in environmental history. He has established international and multidisciplinary research networks that have initiated systematic studies of the development of environmental politics, science and the technology in the Baltic Sea Region.

Tina Löffelsend is adviser for climate and energy policy for the Trade Union Foundation Social Society – Sustainable Development in Berlin. She earned an MSc in political science from the Free University Berlin and previously worked in the foreign policy field at the European and German national parliament.

Tuomas Räsänen is PhD researcher in the Department of General History at the University of Turku, Finland. His doctoral thesis examines ecosystem changes in the marine environment and the history of Baltic Sea protection during the third quarter of the 20th century.

Riku Varjopuro is a senior researcher at the Finnish Environment Institute's Research Programme for Environmental Policy. His main research interest is interactions between participants of social and political processes of natural resource management – for instance, fisheries management, river basin management and biodiversity conservation.

Acknowledgements

Launching a project with a few old colleagues, a handful of fresh doctoral students and numerous new faces is always more or less like a journey into an unknown territory. To be able to undertake this kind of journey and achieve desired results, it is essential to have backup, resources and good guidance, as well as friends and colleagues with whom to talk.

We were lucky to have sufficient resources, good guidance and helpful friends for this book and for the entire Governing the Common Sea (GOVCOM) research project within the Finnish Baltic Sea Research Programme (BIREME). We are grateful, first, to the Academy of Finland for funding and supporting our project so generously. As coordinators, Dr Kaisa Kononen and Dr Tuula Aarnio were very involved and were particularly good at encouraging and motivating both us and the teams working on other projects in the BIREME research programme.

We received significant assistance during the preparatory phase of the project from several colleagues in our universities and institutions. We would like to mention, in particular, Professor Riley Dunlap, who was based at Åbo Akademi University, Finland, during this time for his helpful comments. During its implementation, the project also received institutional guidance from an international advisory board comprising Professor Leonardas Rinkevicius, Professor Erik Bonsdorff and the late MEP Matti Wuori. During the writing of the book we also received valuable referee comments from Professor Bob Evans and Dr Kate Theobald. The City of Turku and Mikko Jokinen, in particular, were very supportive when we organized an assessment meeting with the advisory board. We would also like to thank the board for the environmental research carried out at Åbo Akademi University and, in particular, Professor Tage Kurtén as chair and Ea Blomqvist as coordinator for their active support of interdisciplinary environmental research.

Valuable practical help during the preparation of the application and during the compilation of the manuscript was received from several of our colleagues and doctoral students at Åbo Akademi University, at the Social Science Centre Berlin and at the University of Greifswald: Edward Cameron, Pamela Dorsch, Linda Kaseva, Dr Lise-Lotte Lindfelt, Dr Krister Lundell and Maria Sjöblom. One of the most important ingredients in the success of the project was the professional administrative support provided by Annika Stålfors.

In addition to the editors, our project team included Björn Grönholm, Professor Yrjö Haila, Dr Ann-Sofie Hermanson, Dr Aija Kettunen, Kati Kuitto, Panu Kontio, Tina Löffelsend, Tea Nõmmann and Riku Varjopuro. We would like to thank them for their active participation in both BIREME and GOVCOM events. We also received valuable help from another BIREME project – The Sea and Societies – in the form of an article by Dr Simo Laakkonen and Tuomas Räsänen, which filled some gaps not covered by our own project.

Finally, this book would never have been published without the generous support of two bodies at Åbo Akademi University. The steering group for the Interdisciplinary Centre of Excellence - Democracy: A Citizen Perspective and the publishing board for the Åbo Akademi Foundation both provided financial support for the publication of this book.

Since ours was a long project with an extensive network of contacts, it is virtually impossible to remember and acknowledge all support that we received during recent years. We can only hope that we have not omitted anyone who contributed to our endeavours.

Marko Joas, Kristine Kern and Detlef Jahn
Åbo, Stockholm and Greifswald,
January 2008

Preface

Why **BIREME?**

Some 90 million people in nine countries live in the Baltic Sea Region. The Baltic has long been an important region both in economic and political terms and therefore a focus of major interest. The nations and states that inhabit the region have shared and experienced many historical events – most recently, the joining of new member states to the European Union in May 2004. This has made the Baltic Sea a genuinely common European sea since eight out of the nine Baltic Sea countries are now members of the European Union.

The Baltic Sea suffers from several environmental problems caused by human activities and the complexity of the natural environment. These growing problems are receiving increased attention both nationally and internationally. In Finland, scientists have been concerned about the poor state of the Baltic Sea's environment. During the late 1990s, with the frequent occurrence of large rafts of blue-green algal blooms, concern about the environmental state of the Baltic Sea also increased among decision-makers and the general public. The Programme for the Protection of the Baltic Sea was adopted by the Finnish parliament in 2002. There was a mutual understanding that gaps existed in the scientific knowledge needed for conducting management actions for the protection of the Baltic Sea. The initiative to set up a Baltic Sea research programme originally came from the scientific community, and in 2001 the board of the Academy of Finland decided to launch the Baltic Sea Research Programme, BIREME (see www.aka.fi/bireme) for 2003 to 2005. BIREME was co-funded with a total of 5.88 million Euros by the Finnish Ministry of the Environment, the Ministry of Agriculture and Forestry, the Ministry of Transport and Communications, the Maj and Tor Nessling Foundation and the Russian Foundation for Basic Research.

BIREME's objectives were to deepen our understanding of the conditions for science-based management of environmental issues in the Baltic Sea. The programme focused (and invited research proposals) on research aiming to combat problems caused by eutrophication and harmful substances, as well as on the maintenance of biodiversity and the sustainable use of marine resources. The specific themes of the programme were:

- analysis of change in the Baltic Sea and its drainage basin;
- interactions between the land, coast, air and open sea; and
- societal and environmental interactions in the Baltic Sea Region.

Human impact upon the Baltic Sea and its relation to the economic and social functions of society were acknowledged, and interdisciplinary research linking these activities was encouraged. An analysis of changes to economies and institutions in the Baltic Sea states was considered important in order to assess the predictability of change and to gain insight into future developments.

The final content of BIREME was formulated on the basis of the submitted, successful project proposals, the 24 projects chosen in six thematic sectors: eutrophication; fish biology and fisheries; winter ecology; history and governance; biodiversity; and toxins.

GOVCOM IN BIREME

In order to be successful in producing answers and practical solutions to improving environmental conditions, Baltic Sea research needs an interdisciplinary approach and strong dialogue across the traditional disciplinary boundaries. However, an interdisciplinary approach in research is young, and established traditions, including research training, are still lacking.

Within BIREME, the Governing the Common Sea (GOVCOM) project – one of the few projects representing non-natural sciences – faced major challenges in building bridges and dialogue between different disciplines, that is, between the natural sciences and social sciences. In joint BIREME events, such as annual symposia and workshops, as well as in publications, the GOVCOM project has been actively initiating and maintaining dialogue between disciplines. As the outcomes of the BIREME programme have now been evaluated, the role of GOVCOM is acknowledged as a forerunner in opening pathways to establishing collaboration, which will lead to true interdisciplinary research actions.

The Baltic Sea is our common sea; environmental threats are our common problem. Hence, research efforts also need to involve both national and international cooperation at all levels. Accordingly, one of the aims of the BIREME programme was to foster international research collaboration. This book is a tangible and successful illustration of international teamwork. As a BIREME project, GOVCOM pulled more than its weight in producing this book, in collaboration with German research groups, as a valid document on Baltic Sea Region governance. This book definitely deserves to be at the top of the reading list for all those working for a better future for the Baltic Sea Region.

Tuula Aarnio
BIREME programme manager
Academy of Finland
January 2008

List of Acronyms and Abbreviations

AFLRA	Association of Finnish Local and Regional Authorities
ANP	Association for Nature Preservation
BIREME	Finnish Baltic Sea Research Programme
BSEP	Black Sea Environmental Programme
BSNN	Black Sea NGO Network
CBSS	Council of Baltic Sea States
CCB	Coalition Clean Baltic
CEE	Central and Eastern European
CH_4	methane
CIS	Commonwealth of Independent States
CO	carbon monoxide
CO_2	carbon dioxide
CSCE	Conference on Security and Cooperation in Europe
DDT	dichlorodiphenyltrichloroethane
DF	decoupling factor
DFG	German Research Foundation
EAP	Environmental Action Programme
EBRD	European Bank for Reconstruction and Development
ECE	United Nations Economic Commission for Europe
EIA	environmental impact assessment
ELF	Estonian Fund for Nature
ENP	European Neighbourhood Policy
EnvCom	Commission on Environment (*of* UBC)
EPR	environmental performance review
ERC	European Research Council
ESDP	European Spatial Development Perspective
EU	European Union
EVS	European Values Survey
FAO	Food and Agriculture Organization of the United Nations
FAOSTAT	Food and Agriculture Organization Corporate Statistical Database
FIFG	Financial Instrument for Fisheries Guidance
FRG	Federal Republic of Germany

GDP   gross domestic product
GDR   German Democratic Republic
GEF   Global Environment Facility
GiK   *Gråsälen i Kvarken* (Grey Seal in Kvarken project)
GOVCOM   Governing the Common Sea project
HELCOM   Helsinki Commission (Baltic Marine Environment Protection
                Commission, governing body of the Helsinki Convention)
Helsinki Convention   Convention on the Protection of the Marine Environment
                of the Baltic Sea Area
IBSFC   International Baltic Sea Fishery Commission
ICLEI   Local Governments for Sustainability
IJC   International Joint Commission
IMF   International Monetary Fund
INTERREG   a Community initiative that aims to stimulate interregional coopera-
                tion in the EU, financed under the European Regional Development
                Fund (ERDF)
IUCN   World Conservation Union
JA   joint action
JCP   Baltic Sea Joint Comprehensive Environmental Action Programme
KLIMP   Climate Investment Programme funding scheme in Sweden
LA21   Local Agenda 21
LASALA   Local Authorities Self-Assessment of Local Agenda 21 – project
LEP   Law on Environmental Protection (Lithuania)
LES   Lithuanian Environmental Strategy
LGM   Lithuanian Green Movement
LIFE   the EU financial instrument supporting environmental and nature
                conservation projects throughout the EU, as well as in some cand-
                idate, acceding and neighbouring countries
LIP   Local Investment Programme for sustainable development in
                Sweden
LVRLAC   Lake Victoria Region Local Authorities Cooperation
MARE   Maritime Research on Eutrophication
MCSD   Mediterranean Commission on Sustainable Development
MDSD   Most Different System Design
MEP   Member of European Parliament
MEPRD   Ministry of Environmental Protection and Regional Development
                (Latvia)
MFA   Swedish Ministry for Foreign Affairs
MLG   multilevel governance
MSSD   Most Similar System Design
n   total population sample size
NAP   Nordic Action Plan
NATO   North Atlantic Treaty Organization

NCSD	National Commission on Sustainable Development (Estonia)
NDEP	Northern Dimension Environmental Partnership
NEAP	National Environmental Action Plan (Estonia)
NEPP	National Environmental Policy Plan (Latvia)
NEPP	National Environmental Protection Programme (Lithuania)
NES	National Environmental Strategy (Estonia)
NGO	non-governmental organization
NIB	Nordic Investment Bank
NO_x	nitrogen oxide
NPAA	National Programme for the Adoption of the Acquis (Estonia)
OECD	Organisation for Economic Co-operation and Development
OILPOL	International Convention for the Prevention of Pollution of the Sea by Oil
OSPAR Convention	Convention for the Protection of the Marine Environment of the North-East Atlantic
PCB	polychlorinated biphenyl
PHARE	'Poland and Hungary: Aid for Restructuring of the Economics' – a pre-accession instrument financed by the EU that also assists other applicant countries of Central and Eastern Europe in their preparations for joining the EU
PITF	Programme Implementation Task Force
SEA	Single European Act
SEA	strategic environmental assessment
SNE	seconded national expert
SO_x	sulphur oxide
SOG	Senior Officials Group
TACIS	'Technical Aid to the Commonwealth of Independent States' – an EU funding instrument
TBestC	Transferring Best Environmental Solutions between Towns and Cities
UBC	Union of the Baltic Cities
UK	United Kingdom
UNCED	United Nations Conference on Environment and Development
UNEP	United Nations Environment Programme
US	United States
VASAB 2010	Vision and Strategies around the Baltic 2010
VOC	volatile organic compound
WTP	willingness to pay
WVS	World Values Survey
WWF	World Wide Fund for Nature

Part I

Introduction

Governance in the Baltic Sea Region: Balancing States, Cities and People

Marko Joas, Detlef Jahn and Kristine Kern

INTRODUCTION

Most of the contemporary research on issues of environmental governance suggests that the whole system of government in the arena of environmental policy-making is in a phase of change, as national governments are under stress from several new political agents. Theoretical debates and empirical studies indicate that, in addition to the traditional nation state-centred policy-making system (which includes international cooperation), political power is also being exercised to an increasing extent at the transnational and local levels of society. A simultaneous movement of political power is occurring upwards to transnational levels of government and downwards to local communities. Sub-national units such as local governments, civic organizations and even loosely constructed networks introduce their own environmental policies. Global sustainability problems are created by the interaction of all societal levels, and a new politics of sustainability involving local, national and regional, as well as global efforts must be implemented to solve these problems. However, even if the nation state handed over some tasks to supra- and transnational or regional actors (Zürn, 1998), and even if the interaction between national, international and regional arenas became more complex than before (Frieden and Martin, 2002), national governments remain important players in the political process. They have also responded to this new situation by introducing programmes promoting ecological modernization as well as new policy instruments that involve local communities and other actors as citizens and stakeholder organizations.

The Baltic Sea Region is of special concern from both an environmental point of view as well as from a governance perspective. The region may be considered as a microcosm for a wider Europe as a place where West and East come together. This means that the Baltic Sea Region is an ideal test case to investigate the compatibility of economic and environmental concerns. As in the rest of Europe, the division between the Western and the Eastern part of the Baltic Sea Region is characterized by tremendous inequality (Jahn and Werz, 2002). This has the potential to cause discussion and distrust rather than unity of purpose; therefore, it is essential to harmonize the development of these different levels of wealth. Such economic adjustment has, of course, environmental consequences. For a long time, wealth creation has been associated with environmental degradation. Even if this relationship has turned in most Western societies and economic improvement is no longer necessarily connected to environmental problems – some even argue it has the opposite effect – the potential for the Eastern countries to develop to the level of their Western neighbours without damaging environmental impact remains open to debate. The balance of economic and environmental concerns is an imperative for a unified Europe and for the Baltic Sea Region, where environmental concerns already had a high priority during the Cold War.

There are several reasons why environmental concerns traditionally have a high saliency in the Baltic Sea Region. The Baltic Sea itself is highly vulnerable to pollution, which has already led to transnational cooperation. At the same time, the region is an ideal setting for our research because it has introduced several new forums for sustainable decision-making, while showing considerable strength in existing administrative and political structures. These changes are the prime focus of this book.

FROM GOVERNMENT TO GOVERNANCE: SOME THEORETICAL NOTES

Policy-making has traditionally been seen as a top-down system, from the international level down to the local one. From this perspective, nation states are traditionally considered to be both autonomous and powerful. The Baltic Sea Region contains states at very different levels of economic, administrative and political development. This longstanding pattern has, however, been under serious stress during the last decade. In several empirical and theoretical studies, the central position of nation states has been questioned. We are witnessing at least a partial 'hollowing-out' of the nation state as a political authority, or, at the very least, a new distribution of tasks between different agents. This is particularly true with regard to environmental and sustainability policies. A simultaneous movement of political power is occurring up to transnational levels of government and down to local communities. Sub-national units, including local authorities, civic organizations and even loosely constructed networks, introduce their own policies (see, for example, Rhodes, 1996; Dryzek, 1997; Gibbs, 2000; Evans et al, 2005). Global sustainability problems are seen as stemming from the interaction of all societal levels, and a new politics of sustainability involving local, national, regional and global efforts must be implemented to ameliorate them. National governments have responded to this situation by introducing new elements into structures aimed at ecological modernization, as well as new policy instruments that involve a larger share of local communities and other actors (see, for example, Jänicke and Weidner, 1997; Sairinen, 2000).

While becoming a catchword within social sciences, the conceptualization of 'governance' has changed considerably during the last 15 years. It has several new meanings, some of which simply highlight different parts of the same phenomenon, while others describe altogether different phenomena. 'Governance' is used both as a concept describing the empirical efforts by governments to adapt themselves to changes in their external environments, but also as a (pre-) theoretical concept highlighting the changing role of the state in the process of coordinating different social systems (Pierre, 2000, p3).

Rhodes (2000, pp56–61) and Pierre and Peters (2000, p14) highlight several different definitions used by researchers and research teams. Governance can either be seen within the boundaries of business administration (discussed under the theoretical heading of 'corporate governance') as a way of introducing business methods to the public sector (in literature highlighting 'new public management' methods); as a guidebook for rules of conduct within administration, often promoted by international monetary agents (called 'good governance'); as a change in the way that international cooperation patterns emerge; as 'international interdependence' between different agents; as the loss of government power to other actors ('socio-cybernetic system'); as a change in the economic conditions for society as a whole ('new political economy'); and as a change in the contacts between political actors, states, local government ('policy network theory') and public–private partnerships. These approaches are all very different in content and in the way in which academics have approached the debate on normative, empirical or theoretical levels.

Political science tends to define governance as a change in the processes of interaction between different political actors, despite the political and administrative level in focus at that time. Within this context phenomena such as multilevel governance, the hollowing-out of the nation state process and various forms and levels of networking, all stress changes both in terms of who is part of the political process, and how this process is conducted today. The change is foremost about an 'erosion of traditional bases of political power', although this can, notwithstanding, be of several different forms (Pierre, 2000, p1).

There is, however, a scholarly controversy on the level of this erosion. Lundqvist (2004, p19; see also Jahn, 2006) presents two opposing views on the impact of this change. On the one hand, he argues that the hierarchical relations defining traditional government are still in place and are highly functional. Governments still establish the basic rules of the game by which other players need to abide. However, even this view acknowledges the fact that to some extent the role of the government is changing 'from a role based in constitutional powers towards a role based in coordination and fusion of public and private interests' (Pierre and Peters, 2000, p25). At the other end of the change process:

> ... *central government is seen as dwindling to a position as one among equals within a core structure of governance consisting of inter-linked networks and communities with both public- and private-sector participants, mutually interdependent on each other for resources such as money, expertise and legitimacy. (Lundqvist, 2004, p19)*

The basic change is seen as affecting the position of national governments. Local governments are perceived as gaining in power, widening their political freedom of movement both within the country and, to a higher extent, also outside the country. This means, in practice, that units other than national governments influence the policy processes at the local level, through sub-governmental transnational networks and international organizations. This potentially gives the European Union a further channel to change political behaviour through local government, and this approach has been visible regarding sustainability policies.

On the other hand, the autonomous position of local authorities is constrained by new local-level actors, different from those participating in the 'normal' political process. The scope of the political process has widened and is becoming more open.

We expect that an analysis of multilevel governance phenomena in the Baltic Sea Region will provide valuable data on several grounds. First, the traditional Nordic welfare model that lifts local government into an autonomous, but also responsible, position in its interaction with the public can be defined as a manifestation of governance (Baldersheim and Ståhlberg, 2002). There is, thus, a tradition of governance in the area.

Second, the 'policy process environment' in the Baltic Sea Region seems to be well suited to analyses of multilevel governance since several policy networks with different actors are present in the area. In fact, even if the majority of these actors are rather new, their activities have roots prior to 1990. Local authorities in this region are also active within the networks with the strongest role in this policy field (Kern and Löffelsend, 2004).

Finally, the development of international contacts within the specific policy field of sustainable development seems to be of particular interest for local governments (Joas, 2001, p261).

TOWARDS GOVERNANCE IN THE BALTIC SEA REGION: WHY IS THIS REGION IMPORTANT FOR RESEARCH?

The Baltic Sea Region is of special interest for research because it is widely regarded as a pioneer in the introduction of new modes of governance. There are several reasons for this.

First, this shallow brackish water reserve is highly vulnerable to stress from modern industrialized countries. This problem was acknowledged at an early stage by the countries surrounding the Baltic Sea. Even during the Cold War period, cooperation across the Baltic Sea was comparatively strong, particularly in the area of environmental policy. International efforts were conducted, for example, through the Helsinki Convention on the Protection of the Marine Environment of the Baltic Sea Area (HELCOM), which was signed by all countries surrounding the Baltic Sea in 1974 just after the institutionalization of environmental policies in the Nordic countries. It became the first international regime on the protection of a regional sea and served as a model for many other regions in the world. Yet, it is obvious that the efforts based on this traditional form of international cooperation and coordination have proven inadequate so far – both nationally and internationally. While some positive changes are visible in the Baltic Sea, other indicators suggest deterioration. Thus, the need for additional efforts is clear, including new instruments and new actors.

Second, there is a tradition of general international cooperation in the region. The rapid development of various new forms of international, intergovernmental and transnational governance in the Baltic Sea Region was triggered mainly by three events:

1 the end of the Cold War and the transition of the former state socialist countries to market economies;
2 the United Nations Conference on Environment and Development (UNCED) in Rio de Janeiro where Agenda 21 was signed in 1992;
3 the enlargement of the European Union, with a first wave in 1995 (Sweden and Finland) and a second wave in 2004 (Poland, Lithuania, Latvia and Estonia).

All three events represent important milestones and drivers that set the region apart from other parts of the world.

The region experienced a major political change just a decade ago, introducing new elements into the environmental political arena. When the Cold War ended, the Baltic Sea Region developed a highly dynamic area of cross-border cooperation. New transnational networks, such as the Union of Baltic Cities, emerged. While multi-stakeholder approaches, such as Baltic 21, include governmental actors as one of the major actor groups, among others, transnational networks are a new governance type that are based on networks of civil society and sub-national actors. The Baltic Sea Region appears to be a fertile ground for the emergence of transnational networks. In this region, the dynamics towards the emergence of such networks seem to be stronger than in other European regions. This process started immediately after the fall of the Iron Curtain, and many transnational networks emerged from the 'bottom up' and independently from nation states. Indeed, it is possible that the Baltic Sea Region is becoming a model for the future of a unified

Europe, replacing the East–West division. However, this can only be successful if the economic adjustment of the four Eastern riparian states is accompanied by a similar development in the environmental sphere.

Further, the Rio conference in 1992 influenced developments towards sustainability. In Rio, new approaches were discussed and adopted, among them new forms of citizen and multi-stakeholder participation. These developments affected not only the individual countries in the Baltic Sea Region, but also the region as a whole. New trends in international and sub-national involvement in sustainability politics seem to be common throughout the region. As a result of the Rio conference, numerous Agenda 21 processes were launched at national and local level. The Nordic countries became pioneers regarding the implementation of Local Agenda 21 (LA21). In addition, an integrated Agenda 21 process for the Baltic Sea Region, Baltic 21, was initiated by the Council of Baltic Sea States in 1996. It became the world's first regional Agenda 21 process.

Finally, it must also be taken into account that governance in the Baltic Sea Region is now embedded within European governance and will lead to the Europeanization of the Baltic Sea Area. With European Union (EU) enlargement, the Baltic Sea is surrounded only by EU member states, with the sole exception of Russia. Therefore, European integration appears to offer a real chance to clean up the Baltic Sea.

The Europeanization of the Baltic Sea Region gathered momentum when Sweden and Finland became EU members in 1995, and another wave of integration and convergence started when the European Union agreed on its Eastern enlargement. It is evident that the different stages of political and democratic development among nations within the region play an important role in the selection and achievement of environmental goals in the countries surrounding the Baltic Sea; but these different stages of development can stimulate new forms of cooperation and learning at the national as well as at the sub-national level.

Pre-accession processes enhanced and accelerated cooperation among the countries in the Baltic Sea Region. Given that the Baltic Sea can be considered as a link between EU member states and the accession countries in Eastern Europe, these trends are very important for the enlargement process. Enlargement has resulted in a fundamental change of governance in the Baltic Sea Region, although the candidate countries in the region were already following the lead of Brussels and EU politics in the pre-accession phase. After EU enlargement, governance of the Baltic Sea Region becomes increasingly embedded in the supranational multilevel structure of the EU. State-centred governance is replaced by multilevel governance that causes political actors to interact across different levels of government. It is already evident that many governmental and non-governmental actors in the Baltic Sea Region orientate themselves towards the EU instead of only acting within their own country.

RESEARCH QUESTIONS

The main objectives of this book are to:

- deepen understanding of the origins, development and operation of traditional environmental governance in the Baltic Sea Region;
- examine newly introduced methods of governance in the region, highlighting developments such as the new participatory approaches;
- analyse the effects of some of the changing patterns of governance on local actors, local communities and the state of the local environment; and
- provide a synthesis of the patterns of change we discover in governance modes in the Baltic Sea Region.

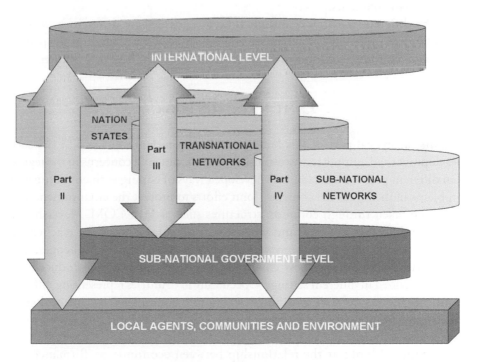

Figure 1.1 *The research model*

The principal driver of these research questions is to produce a better understanding of how traditional forms of environmental governance interplay with the new forms of governance that are already visible in the Baltic Sea Region. The Baltic Sea Region is special in this regard, having introduced several new forums for sustainable decision-making, while at the same time showing considerable strength in already existing structures. One of the most interesting aspects of contemporary environmental policy-making is that it is no longer simply a 'top-down' process, as was the case historically. Participatory approaches have been analysed and evaluated several times in local settings, but as far as we know, never with a transnational effort in mind. Attention will be given to potential differences in this regard between the newly democratic countries and the other nations in the Baltic Sea Region.

Today we face new forms of governance emerging from lower levels, as well as upper (national and international) governmental levels. Although stemming from international events, tools such as Local Agenda 21 are explicitly designed to encourage local levels of government to play key roles in environmental policy-making. In contrast, other new tools are being promoted as tools that national governments can employ to guide decision-making to achieve sustainability. We are additionally interested in the effects of these political changes on local communities and local environments.

The different research questions therefore investigate different aspects of governance as a phenomenon in order to produce a holistic analysis of the pattern of change regarding environmental governance modes in the Baltic Sea Region.

OUTLINE OF THE BOOK

The first analytical part of this book provides an in-depth analysis of the early development and administrative evolution of environmental concerns in the region. Ten different political systems have developed ways of coping with environmental problems in their national contexts. Joint efforts to protect the environment have also been taken through common structures such as HELCOM. Nevertheless, there are differences in how and when environmental issues have reached the political agenda, and how the systems have responded to new challenges. This part of the book examines how environmental issues have been introduced and institutionalized in different settings – specifically in countries representing different political cultures, as well as different economic circumstances and ecological challenges. An international dimension on how HELCOM was introduced to the area is also included.

Chapter 2 looks at the relationship between economic performance and environmental impact in countries of the Baltic Sea Region. A macro-structural approach is used to analyse Finland, Sweden, Denmark and Germany over a period of almost 50 years, as well as Poland, Lithuania, Latvia and Estonia over a shorter period of 10 years. The results prove complex. The patterns of development of economic growth and environmental pollution in the Eastern riparian countries do not seem to follow the trend of Western riparian countries of the Baltic Sea and Western highly industrialized countries, in general, inasmuch as the decoupling of economic growth and environmental pollution occur at a much earlier stage of the economic development in the East.

Chapter 3, which is based on another BIREME project, The Sea and Societies: Approaches to the Environmental History of the Baltic Sea, examines the role of Finland in the process that led to the signing of the Helsinki Convention from the perspective of international politics. The objective is to demonstrate that the convention not only aimed to protect the environment, but also had a political dimension that determined the framework for the drafting and the contents of the convention. It is true that the negotiations that led to the signing of the Helsinki Convention could not have progressed if environmental pollution had not been a real issue; but environmental factors were not the only reason for the eventual signing of the convention. The perspectives of environmental protection alone cannot explain how the Helsinki Convention came to be concluded even though Northern Europe was sharply divided by the Cold War at the time.

The institutionalization process at a national level is further analysed in Chapter 4. All of the advanced industrial societies have experienced a shift towards more post-materialist value orientations. A widespread change in attitudes brought environmental issues to the political agenda during the early 1970s. Movements, gathering around different issues and influencing decisions on the local level, became more organized and challenged national political decision-making. Nordic countries witnessed a particular and early interest in environmental issues. The focus of this chapter is the institutionalization of environmental concerns in different settings – in Nordic and Baltic countries – representing different political cultures, as well as different economic circumstances and ecological challenges. The results are examined by nations since this is the traditional arena for actions aimed at environmental policy-making.

This theme is further explored in Chapter 5, where the process of environmental institution-building in the Baltic republics is discussed. Transition is a key element in influencing environmental policy-making, and society as a whole in the Baltic states is in transition. The time frame for changes is different compared to Western European and Nordic countries, where environmental policy and governance developed gradually in tandem with increasing environmental awareness over decades. The purpose of this chapter is to briefly describe the development of environmental administration and environmental policy in the Baltic states and to understand some of the patterns within this development. Reference is made to the development of environmental policy and governance in Western European and Nordic countries. However, it would not be justified to try to use the Nordic consensus (corporative) model to describe and analyse development in the Baltic republics since the starting point and the circumstances have been so different.

Part III of the book highlights some of the new structures of governance and examines the introductory phases of new forms of environmental policy that have been introduced to the Baltic Sea Region. The central question is how political innovations such as Local Agenda 21 are introduced and adopted among governmental entities within the region. The Baltic Sea Region shows considerable variation in environmental policy among nations due to historical and contemporary differences in state regimes, administrative styles and political cultures. Yet, ten independent nations are connected by a common sea, and despite different national circumstances and needs they are stimulated to work together to protect that sea and to abide by international agreements and conventions. The situation is made even more complicated by new forms of environmental governance that involve an increasing policy-making role for municipalities and other local and regional governments, as well as non-governmental organizations (NGOs) and networks.

After the end of the Cold War, the Baltic Sea Region developed into a highly dynamic area of cross-border cooperation and transnational networking. This process is highlighted on a general level in Chapter 6. Various types of governance beyond the nation state are discussed here, ranging from international regimes, such as the Helsinki Convention for the Protection of the Baltic Sea, to transnational networks, such as the Union of the Baltic Cities. Governance towards sustainable development of the Baltic Sea Region undoubtedly requires national governance beyond the nation state. In this respect, transnational networks and the European Union provide promising new approaches that can complement the traditional forms of international and intergovernmental cooperation between nation states. These new governance types represent two parallel trends: first, a development towards transnationalization; and, second, the Europeanization of the Baltic Sea Region.

A local dimension to this development is brought forward in Chapter 7. Local Agenda 21 has proven to be a success both within the context of the Eastern European countries and within the older EU member states, and new structures have fairly rapidly been diffused over the former geopolitical borderlines. This chapter asks if this development really is unique or is just a result of good marketing. The two groups of Central and Eastern European (CEE) countries, Baltic region CEE countries and other CEE countries are compared with each other. It is assumed that the emerging new multilevel governance structures and extensive use of networking in the region can at least partly be seen as an explanation for the possible activity difference. This assumption is tested with a comparative database from 138 European cities active in LA21 or similar sustainability policy process. This Local Authorities Self-Assessment of Local Agenda 21 (LASALA) database is collected from self-evaluations by cities all over Europe.

Part IV, the final empirical part of the book, takes up the question of how interactions between regional governance for conservation policy and policies to sustain local livelihood can be combined for mutual benefit. The chapter deals with the impacts of changing national and international environmental policies on local community livelihoods and local environmental governance. The analysis is focused on the interaction between nature conservation and local use of resources, with a special emphasis on the development of forms of governance dealing with the interaction. The research sheds light on the transformation and 'translation' of national and international conservation policies onto the local level. The success or failure of conservation policy depends largely upon its success in modifying local practices and inspiring new forms of regional governance, and interactions with local socio-economic systems are essential in this respect. In practical terms, the results will help to:

- understand the dynamics of multi-scale governance in conservation; and
- develop governance models for mitigating local and regional conservation conflicts.

These aspects are highlighted in Chapter 8: the policy goals of nature conservation are today often formulated at the international level through laws and conventions. National legislation adopts these international decisions and the states often organize the actual implementation (planning and enforcement) of conservation policies in a hierarchical way. A consequence of the implementation of nature conservation policy is that it influences human activities at the local level in a substantial way. Difficult conflicts arise when conservation policy requires the restriction of traditional local activities. Incompatible interests in environmental resources result in environmental conflict. This chapter aims at describing what has happened in the Kvarken area in the Northern Baltic Sea regarding the conflict between grey seal conservation policies and local fisheries. The chapter describes regional actors' own experiences and provides a description of the project. The results are related to a more general discussion on the governance of complex multilevel issues.

The empirical discussion in the book is concluded in Chapter 9, which discusses a substantial transformation that occurred in nature conservation in contemporary society during the last decades of the 20th century. This transformation occurred in tandem with the surge of new environmental problems within the public consciousness since 1960. As a consequence, environmental policy was established as a new sector of public policy. In a formal sense, the unification of environmental policy is a goal worth pursuing: unification reduces ambiguity in the criteria used to assess environmental problems and to evaluate success versus failure in addressing them, and it also helps to integrate environmental goals within other sectors of public policy. However, there is another side to the coin. The field of environmental policy is made up of a highly heterogeneous set of specific problems. At some level of resolution, differentiation between policy instruments and between ways of implementation is necessary, and it is not *a priori* obvious what this level is.

In the context of Chapter 9 (and this book as a whole), however, a comprehensive review of the Baltic region's environmental problems is not possible. Chapter 9 uses three recent articles published in the Finnish journal *Vesitalous* (issue 2, 2006) as primary source material. These articles summarize results of major research projects funded by BIREME. They are therefore representative enough for the purposes of this analysis. However, they are almost exclusively based on the natural sciences. To complement the setting, Yrjö Haila brings in two additional case studies on the social dimensions of environmental conflicts along the Finnish Baltic coast: on seal conservation in Kvarken, the Bay of Bothnia, and aquaculture versus eutrophication in the Archipelago Sea. In organizing and analysing these materials, Haila makes use of the notion of 'problem space', constructed as an analogy with physical space. The last sections of the chapter return to the question of problem closure and environmental governance.

In Part V, Chapter 10 combines the results of the individual chapters and offers a synthesis of environmental governance concerning the Baltic Sea Region. The focus is on new forms of governance and the factors that influence the introduction of cooperation and integration of environmental policies in different countries and at different societal levels. Environmental politics and policy, as well as the adoption of new forms of governance on different levels, can be seen as a result of existing traditional environmental administrations, public environmental awareness, transnational networks and cooperation, socio-economic and cultural factors, as well as the environmental condition of the Baltic Sea. This synthesis produces a deep understanding of mechanisms for encouraging effective (and preventing ineffective) forms of governance through a transnational and regional effort. In addition, this chapter also takes a look beyond the Baltic Sea Region, providing a preliminary comparison of similar political settings in comparable vulnerable sea and lake areas.

REFERENCES

Baldersheim, H. and Ståhlberg, K. (2002) 'From guided democracy to multi-level governance: Trends in central–local relations in the Nordic countries', *Local Government Studies*, vol 28, pp74–90

Dryzek, J. (1997) *The Politics of the Earth: Environmental Discourses*, Oxford University Press, Oxford

Evans, B., Joas, M., Sundback, S. and Theobald, K. (2005) *Governing Sustainable Cities*, Earthscan, London

Frieden, J. and Martin, L. L. (2002) 'International political economy: Global and domestic interaction', in I. Katznelson and H. V. Milner (eds) *Political Science: The State of the Discipline*, W. W. Norton and Company, New York

Gibbs, D. (2000) 'Ecological modernisation, regional economic development and regional development agencies', *Geoforum*, vol 31, pp9–19

Jahn, D. (2006) 'Globalization as Galton's problem', *International Organization*, vol 60, pp401–431

Jahn, D. and Werz, N. (2002) 'Der Ostseeraum: Eine Zukunftsregion mit ungleichen Voraussetzungen', in D. Jahn, D. and N. Werz (eds) *Politische Systeme und Beziehungen im Ostseeraum*, Olzog, München

Jänicke, M. and Weidner, H. (eds) (1997) *National Environmental Policies: A Comparative Study of Capacity Building*, Springer, Berlin

Joas, M. (2001) *Reflexive Modernisation of the Environmental Administration in Finland: Essays of Institutional and Policy Change Within the Finnish National and Local Environmental Administration*, PhD thesis, Åbo Akademi University Press, Åbo

Kern, K. and Löffelsend, T. (2004) 'Sustainable development in the Baltic Sea Region: Governance beyond the nation state', *Local Environment*, vol 9, no 5, pp451–467

Lundqvist, L. J. (2004) *Sweden and Ecological Governance: Straddling the Fence*, Manchester University Press, Manchester

Pierre, J. (2000) 'Introduction: Understanding governance', in J. Pierre (ed) *Debating Governance Authority, Steering and Democracy*, Oxford University Press, Oxford, pp1–12

Pierre, J. and Peters, B. G. (2000) *Governance, Politics and the State*, St. Martin's Press, New York

Rhodes, R. A. W. (1996) 'The new governance: Governing without government', *Political Studies*, vol 44, no 4, pp652–667

Rhodes, R. A. W. (2000) 'Governance and public administration', in J. Pierre (ed) *Debating Governance Authority, Steering and Democracy*, Oxford University Press, Oxford, pp54–90

Sairinen, R (2000) *Regulatory Reform of Finnish Environmental Policy*, Centre for Urban and Regional Studies Publications A 27, HUT, Espoo

Zürn, M. (1998) *Regieren jenseits des Nationalstaats*, Suhrkamp, Frankfurt am Main

Part II

Introduction and Institutionalization of Environmental Concerns within the Baltic Sea Region

2

Environmental Pollution and Economic Performance in the Baltic Sea Region[1]

Detlef Jahn with Kati Kuitto

INTRODUCTION

Economic growth and environmental performance are crucial for the functioning of modern industrial societies. Sustainable economic growth, in particular, has always been considered a guarantee for stable democracy. During the last three or four decades, however, voices were raised claiming that economic development has to include environmental aspects in order to be sustainable. According to this view, growth without environmental concern is considered to erode societal resources (WCED, 1987, p8). To achieve true environmental sustainability in the Baltic Sea Region, it is essential to reconcile economic imperatives with environmental needs. The wealth of Western nations is historically based on the exploitation of nature. One of the key questions for the future of sustainable development in the Baltic Sea Region is whether the newly democratized riparian countries will achieve similar levels of economic development without abusing the environment in the process. It is therefore essential that environmental governance of the Baltic Sea Region combines the goals of economic growth and environmental protection.

The relationship between economic performance and environmental pollution is not free of tension (Schnaiberg and Gould, 1994), and there are several competing hypotheses about the causal effect of economy on environmental performance (Jänicke et al, 1996). The most popular hypothesis starts from the assumption that economic performance causes environmental problems and increasing economic performance boosts environmental degradation (*prosperity pollution*). The nexus between economic performance and economic growth, on the one hand, and environmental degradation, on the other, was at the centre of the early ecological debate (Milbraith, 1984; Paehlke, 1989).

In this debate it was postulated that there is clear antagonism between economic growth and environmental protection. Later, however, it was argued that not only is there a decoupling of economic growth and pollution, but even that economic prosperity may lead to a cleaner environment (*prosperity cleaning-up*) (Beckerman, 1992). New technologies and a growing service industry are seen as causing less environmental damage than economies centred on traditional production. Richer countries are supposed to develop towards the former and poorer countries towards the latter (Cleveland and Ruth, 1998). However, in the former Communist states, there might also be another kind of relationship in which economic performance and environmental pollution are reduced due to severe economic problems (Jahn, 2003). This *misery cleaning-up hypothesis* would also lead to a statistically positive correlation between gross domestic product (GDP) and carbon dioxide (CO_2) emissions, but would result from other less desirable reasons.

The two former hypotheses (prosperity pollution and prosperity cleaning-up) may imply that we have to deal with a curve–linear relationship. Its turning point will be reached once a certain level of economic performance is achieved. In economics this relationship has been discussed in terms of the Kuznets curve. Simon Kuznets (1955) postulates that income inequality increases and then decreases during the process of industrial development. In the field of environmental studies, this means that up to a certain level there is a clear positive relationship between economic growth and environmental pollution, and after that turning point the relationship reverses itself (Selden and Song, 1994; Grossman and Krueger, 1995; Mátyás et al, 1997; Andreoni and Levinson, 1998; Ekins, 2000, Chapter 7; Harbaugh et al, 2002).

Most of our knowledge regarding relations between economic performance and environmental pollution is based on studies of the richest industrial societies. Even in this context, however, results are mixed (Jahn, 1998). Jänicke and colleagues (1996) conclude that emissions of nitrogen oxides, volatile organic compounds and the amount of municipal waste confirm the prosperity pollution hypothesis in Western industrial societies. Sulphur oxide emissions, major protected areas and waste water treatment confirm the prosperity cleaning-up hypothesis and carbon dioxide emissions and the use of fertilizer show a Kuznets curve relationship. Lyle Scruggs (2003, pp56–62) concludes that there is strong evidence for a Kuznets curve development for an aggregated environmental performance index for 17 highly advanced Organisation for Economic Co-operation and Development (OECD) countries, although in the opposite direction: poor and rich countries pollute more than those nations with intermediate wealth. Where are the riparian countries of the Baltic Sea Region in this context? And is there currently a clear distinction between Western and Eastern nations?

The rationale of this chapter is to analyse the development of the relationship of economic growth and environmental pollution in countries of the Baltic Sea Region by selecting two important pollutants that, according to Jänicke and Weidner (1997), exhibit the reversed U-shape relationship in rich Western European industrial societies: carbon dioxide emissions and fertilizer consumption. CO_2 emissions contribute the largest share of manmade greenhouse gases. These emissions disturb the balance of the Earth's radiative energy budget and may lead to an increase in surface temperature and related effects on climate, sea-level rise and world agriculture (Jäger and O'Riordan, 1996). Nitrates and phosphates contained in fertilizers contribute to a great extent to the nutrient load in aqua-systems and, thus, to eutrophication of rivers and lakes. Besides the domestic impacts of fertilizer use, the eutrophication caused by the nutrient load from agricultural sources is one of the main causes of the disruption of greater aquatic ecosystems. This is especially true for the sensitive maritime ecosystem of the Baltic Sea (Kremser and Schnug, 2002).

The transition countries in the Baltic Sea Region clearly require significant economic growth to increase the prosperity of their respective populations, and to bring them into line with European standards and averages. Environmental degradation may be one consequence of this economic growth. In this regard, the transition economies will likely share this fate with their neighbours in the region and throughout the rest of Europe. As we will see, environmental degradation may not necessarily be an absolute, definitive and certain consequence of economic growth. An analysis of the two environmental pollutants and their relationship to economic growth should provide some indication of whether environmental degradation in the Baltic Sea is inevitable over the coming years.

Which of the competing hypotheses corresponds to the reality in the Baltic Sea Region? Is the development of CO_2 emissions and fertilizer consumption really like a reversed U-shape? How does the development of the relationship between economic growth and environmental pollution in the Eastern riparian countries of Poland, Lithuania, Latvia and Estonia relate to that in the Western riparian countries of Finland, Sweden, Denmark and Germany?

Both CO_2 emissions and fertilizer use are subject to national and international regulatory measures. The reduction of CO_2 gathered momentum on the international stage in the wake of the 1988 Montreal Protocol on Substances that Deplete the Ozone Layer and became a major focus with the 1992 United Nations Conference on Environment and Development (UNCED) at Rio. Modern societies have consequently been aware of CO_2 emissions since the mid 1980s and have intensified their efforts to reduce it during the 1990s. As a consequence, international agreements will have an impact upon a Kuznets curve development.

Countries of the Baltic Sea Region are also committed to reducing environmental pollution from agricultural sources – for example, in the context of the Convention on the Protection of the Marine Environment of the Baltic Sea Region and the European Union Common Agricultural Policy (Ehlers, 2001). Here, too, international agreements can be expected to have an impact upon the development of environmental pollution, at least since the mid 1990s.

ECONOMIC PERFORMANCE, CO_2 POLLUTION AND FERTILIZER CONSUMPTION IN ADVANCED INDUSTRIALIZED COUNTRIES

Jänicke et al (1996, 1997) have compared the level of GDP and CO_2 emissions per capita in 35 countries during 1970 and 1990, as well as fertilizer consumption per capita in 35 countries in 1970 and 1994. The results for CO_2 show a clear positive linear relation for countries with a GDP per capita of less than US$7000. Among these countries are Western nations such as Greece and Portugal, as well as Eastern European states such as Romania, the Soviet Union, Yugoslavia and Poland. A reverse trend can only be seen in Hungary and Czechoslovakia, confirming the prosperity cleaning-up hypothesis with a drop in CO_2 emissions despite increases in the GDP. However, this trend was more common among rich countries such as Sweden, Denmark, Germany, The Netherlands and France, although there were also exceptions, such as Iceland and Norway and, to a certain degree, Switzerland.

A similar trend can be observed for fertilizer consumption per capita. For countries with a GDP per capita of less than US$8000, there is a clear positive linear relation. Most Eastern European countries included in the analysis of Jänicke and Weidner (1997) clearly show this trend, while in richer Western industrialized countries the trend is slightly reversed.

Simple inspection of Figures 2.1 and 2.2 shows that there is a trend towards a Kuznets curve on the cross-sectional level for both pollutants: rich and poor

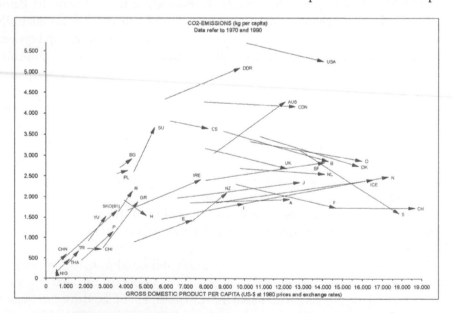

Figure 2.1 *GDP and CO_2 emissions, 1970 and 1990*

Source: Jänicke and Weidner (1997, p318)

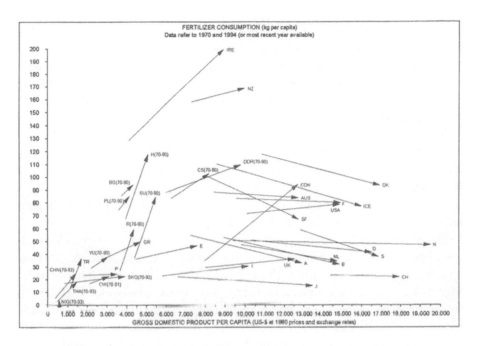

Figure 2.2 *GDP and fertilizer consumption, 1970 and 1994*

Source: Jänicke and Weidner (1997, p319)

countries have relatively low levels of CO_2 emissions and fertilizer consumption, while moderately wealthy countries pollute most. There is also some evidence that this trend might be confirmed when we consider time trends. For poor countries, we identify ascending lines between 1970 and 1990/1994; for rich countries, this trend is reversed. However, focusing on merely two points in time is highly unreliable, and we will therefore address the development of the relationship between GDP and CO_2 emissions and between GDP and fertilizer consumption for the countries in the Baltic Sea Region in greater detail later on.

RESEARCH DESIGN

For the empirical analysis, we included Finland, Sweden, Denmark and Germany, as well as Poland, Lithuania, Latvia and Estonia. The Russian Federation, which plays an important role in Baltic Sea Region pollution, is of great interest for our analysis, but could unfortunately not be included due to lack of reliable data. Russia as a whole is far too large and diverse; therefore, a successful analysis would require data covering the coastal regions that border the Baltic Sea. Unfortunately, no such targeted data is available and we have opted to omit the Russian Federation rather than take the risk of providing a misleading interpretation.

As mentioned above, CO_2 emissions as well as fertilizer consumption have been chosen as clear indicators for environmental pollution that follow a Kuznets curve development over time. Data on CO_2 emissions are taken from the webpage of the Carbon Dioxide Information Analysis Centre (see http://cdiac.ornl.gov/trends/emis/tre_coun.htm, accessed 16 February 2004). The indicator applied in this chapter refers to CO_2 emission rate per capita. Data on CO_2 in most states of the world, supplied by the centre, reach back – in some cases – more than 250 years. Data on fertilizer consumption have been collected from the Corporate Statistical Database (FAOSTAT) of the Food and Agriculture Organization of the United Nations (FAO) on its webpage (see http://faostat.fao.org, accessed 16 February 2004). The indicator refers to total fertilizer consumption in kilograms per square kilometre of arable land.

Measuring environmental pollution only on the basis of two indicators is not, of course, sufficient. For most environmental indicators, annual data are only available since the beginning of 1990s. This is the case for air pollutants such as sulphur oxides, nitrogen oxides, methane and carbon monoxide. Therefore, we cannot produce a time series analysis analogical to the analysis of the relationship between CO_2, fertilizer use and economic growth for these indicators. After controlling for bivariate correlations between CO_2 and sulphur oxides (SO_x), nitrogen oxides (NO_x), methane (CH_4), volatile organic compounds (VOCs) and carbon monoxide (CO), we can nonetheless say that since closely connected to these, CO_2 stands for other air pollution indicators as well.[2] Findings from a factor analysis also confirm this close relationship.

Using fertilizer consumption as an indicator also has drawbacks. Due to geological and biological factors such as soil composition, nutrients included in fertilizers contribute to eutrophication in different ways in various areas. Consequently, using the actual nutrient load in the aqua-systems as an indicator would be a more proper measure of the environmental pressure caused by fertilizers. Although there are several scientific sources of data on nutrient load, the problem is that data are mainly organized per drainage basin, river or point sources, rather than on a country basis.[3]

The independent variable is GDP per capita taken from Angus Maddison's *The World Economy: Historical Statistics* (2003). Maddison offers annual standardized as well as cross-sectional data comparable over time for all cases we consider. His unit of measurement is the 1990 international Geary-Khamis dollar.

Even though the data we use are highly reliable, statistics have to be treated with caution, particularly with regard to the former Communist states. Bearing all of the above mentioned challenges and provisos in mind, it is, nonetheless, reasonable to consider the data upon which the analysis relies as best data available.

The analysis is based on a bivariate analysis, which is certainly not sufficient for a sophisticated analysis of causality. However, we are not primarily interested in the identification of the causal factors of CO_2 emissions (for the analysis of some other aspects, see Jahn, 2001) or of fertilizer consumption. Instead, we are interested in the relationship between GDP and CO_2 emissions and between GDP and fertilizer consumption in different settings. Although more elaborated investigations are needed, this bivariate analysis may be a first step in this new area of research comparing countries not only from Western Europe, but also including countries from Eastern Europe.

CORRELATION BETWEEN GDP, CARBON DIOXIDE EMISSIONS AND FERTILIZER CONSUMPTION IN COUNTRIES OF THE BALTIC SEA REGION

The overall trend

In order to assess whether there is a linear or curve–linear trend between environmental pollution and the wealth of countries, the following analysis will be conducted in two steps. First, we will look at all eight countries over 50- (CO_2 emissions) or 40-year (fertilizer consumption) periods for the Western countries, and over the last 9 and 10 year periods for Eastern countries. This will give us a very rough impression about the overall trend. In the second step we will analyse the development of the correlation between GDP per capita and CO_2 emissions per capita, as well as between GDP per capita and fertilizer consumption per kilometre squared of arable land in each country individually over time. We begin with an overview of development over time across the countries.

A simple test of a linear or curve–linear relationship can be conducted by looking at a linear and curve–linear regression analysis. The curve–linear analysis uses the square of the GDP per capita. A first answer is provided based on which model better suits our data. The results for the relationship between GDP per capita and CO_2 emissions are quite clear. The R square for a linear model is .246, while it is as high as .405 for a curve–linear relation. In Figure 2.3 a linear and curve–linear fitted line illustrates the results very impressively.

Concerning data on fertilizer consumption, we cannot use the German data because they refer to both the Federal Republic of Germany and the German Democratic Republic. Nevertheless, on the basis of this indicator, we have strong evidence that we are dealing with a curve–linear relationship as well. The R square increases from .315 for a linear model to .493 for a curve–linear model. Figure 2.4 also demonstrates this relationship quite clearly.

However, this time-pooled cross-sectional analysis may cover several subtle relationships. In particular, the difference between Western and Eastern states of the Baltic Sea Region is of high interest for our analysis. As a result, we conduct a more fine-grained analysis for the individual countries. We begin with a detailed look at the development of the relation between GDP and CO_2 emissions in the four Western riparian countries of the Baltic Sea, Finland, Sweden, Denmark and Germany, and then proceed with the same analysis for Poland, Lithuania, Latvia and Estonia.

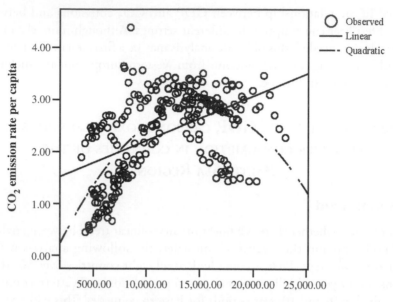

Figure 2.3 *CO_2 emission rate per capita and GDP per capita*

Source: Maddison (2003); Carbon Dioxide Information Analysis Centre

ECONOMIC GROWTH AND CO_2 EMISSIONS

Jänicke and Weidner (1997) based their argument on a cross-sectional analysis of the 35 countries at just two points in time, 1970 and 1990/1994. Although only four Western European countries are included in our analysis, the analysis of Finland, Sweden, Denmark and Germany basically confirms their conclusion, but also adds further important information. Annual time series from 1950 until 2000 – presented in Figure 2.5 – produce some distinctive patterns among the four countries. All countries show a strong positive linear relationship between GDP and CO_2 emissions per capita during the 1950s and 1960s. This period is characterized by strong economic growth and substantial increases in CO_2 emissions. During the 1970s there is an abrupt change and a slow decrease in growth corresponds with lower CO_2 emissions. Even though there is no clear decoupling of GDP and CO_2, the relationship is not as clear cut as it used to be in earlier decades. While the 1950s, 1960s and, to some degree also, the 1970s confirm the prosperity pollution hypothesis, the 1980s support the prosperity cleaning-up hypothesis indicating a reverse U-shape development where increases in both GDP and CO_2 emissions went hand in hand from the 1950s to the 1970s, reached a turning point during

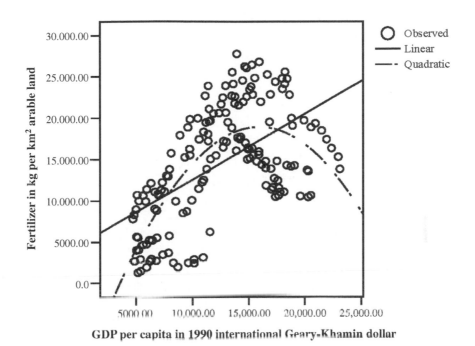

Figure 2.4 *Fertilizer per kilometre squared of arable land and GDP per capita*

Source: Maddison (2003); FAOSTAT

the 1970s, and exhibit a reverse relation (i.e. economic growth and declining CO_2 emissions during the 1980s). However, during the second half of the 1980s and the 1990s the relationship was quite ambiguous among Western riparian countries. Germany and Sweden continued according to the prosperity cleaning-up hypothesis. In Denmark, a somewhat clear trend following the prosperity cleaning-up hypothesis applies only since the second half of the 1990s. In Finland, the positive trend of the first half of the 1980s ceased during the second half of the 1980s, and no clear trend can be identified for the 1990s.

At first glance Sweden appears to be the most impressive example of decoupling prosperity from pollution. While CO_2 levels in Sweden today are comparable to those during the 1950s, GDP levels are three times higher. However, closer inspection reveals that much of this success is based on a substantial reliance on nuclear energy, which (while reducing CO_2) leads to different problems of nuclear waste.

Although less successful in reducing CO_2 emission rates, development in Finland and Denmark during the 1990s does not clearly follow the prosperity pollution hypothesis. In Denmark, recent developments seem to indicate a trend towards a decoupling of economic growth from CO_2 emissions. Finland maintains

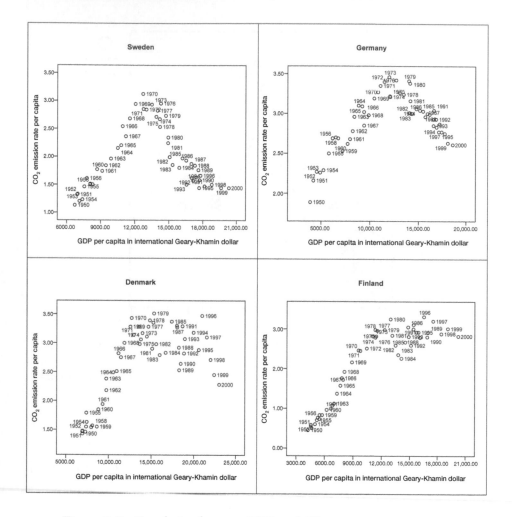

Figure 2.5 *Correlation between GDP and CO$_2$ emissions per capita in Western riparian countries, 1950–2000*

Source: Maddison (2003); Carbon Dioxide Information Analysis Centre

rather high levels of CO$_2$ emissions and GDP growth rates, thus not following the prosperity clean-up hypothesis either.

These findings for the four Western riparian countries of the Baltic Sea are, to a large extent, in line with findings in previous analyses of a more substantial group of Western highly industrialized countries (Jahn, 2003), and support the hypothesis that less wealthy countries (one may identify a baseline level of a turning point around US$12,000) stick to a positive linear relationship between GDP and CO$_2$ emissions, while rich countries follow – at least to a certain degree – the development of a Kuznets curve. Transferring these findings to the Eastern riparian countries of the Baltic Sea as representatives of the transition states of Central and Eastern Europe states suggests a clear positive correlation between GDP and CO_2 emissions, since none of the Eastern European riparian countries of the Baltic Sea has reached the level of US$12,000 so far.

However, the question of whether we are dealing with time effects or wealth remains unanswered. The results suggest a mix of both variables. In all four countries we discover a linear positive relationship between GDP per capita and CO_2 emissions until the first oil crisis in 1973. This clear relationship wanes after that date. However, while we identify a clear reverse trend in Sweden and Germany from the early 1970s, this trend starts in Denmark only during the 1990s. In Finland there is a break of the linear trend during the 1980s; but there is no clear reverse trend in the 1990s. Rather, the relationship between wealth and CO_2 emissions is becoming neutral.

We will now turn to the development of the relation between GDP and CO_2 emissions in the Eastern riparian countries of Poland, Lithuania, Latvia and Estonia. As mentioned above, the prosperity pollution hypothesis may turn into a misery cleaning-up hypothesis in Central and Eastern European countries. This is due to the collapse of most of the industrial sectors following the breakdown of the Communist economic system in these countries, causing unintended and partly drastic positive environmental effects. However, these positive environmental effects are not sustainable since they are based on relative poverty.

Results are given in Figure 2.6. A trend consistent with the misery cleaning-up hypothesis only appears to a limited extent and only for the very first years after the system transition in Poland, Lithuania, Latvia and Estonia. Instead, we observe a trend indicating relatively impressive economic growth during the 1990s, but a continuous decrease of CO_2 emissions. While the development and levels of GDP in these countries equalled those in the Western riparian countries during the 1950s and 1960s, the development of CO_2 emissions was reversed. Hence, there seems to be evidence for the prosperity cleaning-up hypothesis with regard to the four Eastern riparian countries of the Baltic Sea. This is especially true for Latvia and Lithuania, whereas development in the two wealthier Eastern Baltic Sea countries is slightly less clear. A positive CO_2–GDP trend becomes clearer in Poland only during the second half of the 1990s, and in Estonia there is a turn in 1997. Whether this turn indicates a decoupling of the relation between GDP and CO_2 emissions is not yet clear. Estonia has also by far the highest levels of per capita CO_2 emission rates, but also the highest GDP per capita. Altogether, Poland and Estonia, in particular, but also Latvia and Lithuania are among those countries of Central and Eastern Europe with the highest economic growth rates, while still demonstrating a decreasing trend in terms of CO_2 emissions per capita. The results raise some optimism about the relationship between economic and environmental development in Europe over the coming years.

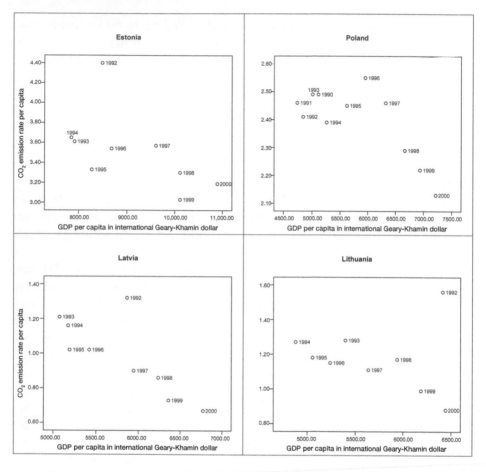

Figure 2.6 *Correlation between GDP and CO$_2$ emissions per capita in Eastern riparian countries, 1991/1992–2000*

Source: Maddison (2003); Carbon Dioxide Information Analysis Centre

ECONOMIC GROWTH AND FERTILIZER CONSUMPTION

We will now take a closer look at the relationship between GDP and the second of our environmental pollution indicators: fertilizer consumption on arable land. In the four Western riparian countries, as shown in Figure 2.7, we observe a similar development in the relationship between GDP and fertilizer consumption. Until 1973/1974, fertilizer consumption clearly increases in line with economic growth. 1973/1974 represents a turning point for Sweden, Denmark and Finland, whereas in Germany the decoupling of GDP from fertilizer consumption takes place around 1979. In Denmark and Sweden, there is a continuous trend of decreasing fertilizer consumption combined with continuous economic growth throughout the 1980s

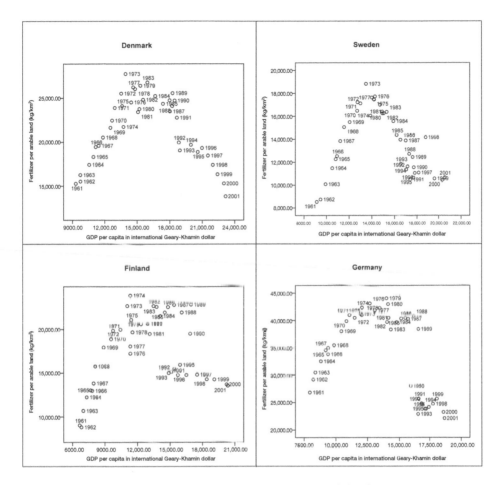

Figure 2.7 *Correlation between GDP per capita and fertilizer consumption per kilometre squared in Western riparian countries, 1961–2001[4]*

Source: Maddison (2003); FAOSTAT

and 1990s. In Finland, the decreasing fertilizer consumption of the mid 1970s is followed by a reappearance of increasing consumption levels during the 1980s. During the 1990s, fertilizer consumption levels off at the quantities of the late 1960s, with a slightly decreasing trend and simultaneous fast economic growth in the second half of the 1990s. In Denmark, Sweden and Germany, the level of fertilizer consumption at the end of the 1990s equals the level of the early 1960s. However, German data are not comparable to the other countries since the data before 1990 are the sum of fertilizer consumption for the Federal Republic and the German Democratic Republic.

All of the Western riparian countries of the Baltic Sea exhibit a clear Kuznets curve development regarding the relation between GDP and fertilizer consumption during the period of 1961 to 2001. We may identify a base line level of a turning point around US$14,000. In the 1970s, there was also a rise of environmental awareness in Western societies, as well as a structural change towards a significant decline of the agricultural sector, both of which might explain the decreasing levels of fertilizer use in our sample countries.

The overall development of the association between GDP and fertilizer consumption thus shows great similarities with the one between GDP and CO_2 emissions in the Western riparian countries, exhibiting a Kuznets curve trend. The opposite seems to be true for the Eastern riparian countries. Results are given in Figure 2.8. The first years of economic breakdown and deterioration of GDP

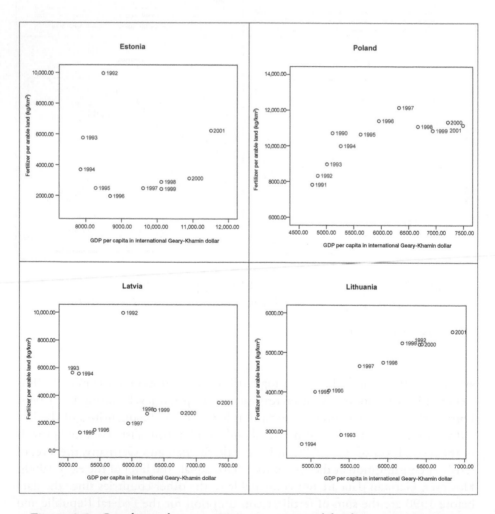

Figure 2.8 *Correlation between GDP per capita and fertilizer consumption per kilometre squared in Eastern riparian countries, 1990/1992–2001*

Source: Maddison (2003); FAOSTAT

witness a drastic decrease in fertilizer consumption in Latvia and Estonia, in particular, and a less drastic decrease in Poland and Lithuania. However, in line with recurring economic growth, fertilizer consumption again increases. This is especially true for Poland, where the period of 1991 to 1997 is characterized by both rapid economic growth and an increase in fertilizer consumption, showing a trend in line with the prosperity pollution hypothesis. Lithuania, Latvia and Estonia also seem to confirm the prosperity pollution hypothesis regarding GDP and fertilizer during the latter half of the 1990s. This trend consequently mirrors development in Western riparian countries during the 1960s. It should be observed, however, that while the level of fertilizer consumption in Latvia and Estonia remains notably lower at the beginning of the 1990s, both Poland and Lithuania currently show higher levels of fertilizer use than at the beginning of the 1990s.

One explanation for increasing levels of fertilizer consumption in the Eastern riparian countries may be found in growing competition within the agricultural sectors of these countries resulting from European Union (EU) membership. However, with the exception of Poland, it needs to be stressed that the level of fertilizer consumption on arable land in Eastern riparian countries is substantially lower than that in Western countries (around 3000 kilograms per square kilometre (kg/km^2) of arable land in 2000 in Estonia and Latvia, and 5000 kg/km^2 in Lithuania compared to 10,000 to 22,000 kg/km^2 at the same point in time in Western riparian countries). The use of fertilizers to increase agricultural yield is widespread across the region. While Western riparian countries account for the greatest share of fertilizer use and the subsequent nutrient pollution of the Baltic Sea catchment area, Eastern riparian countries are currently increasing their use. There is a pressing need for all countries in the region to reconsider their dependence upon these polluting fertilizers. (HELCOM, 2003; see also Turnock, 2002, p71).

DEVELOPMENT OF AIR POLLUTION DURING THE 1990S

As mentioned earlier, we cannot draw very far-reaching conclusions about the relationship between economic growth and environmental pollution on the basis of only two pollutants. However, CO_2 emissions correlate highly with other air pollution indicators, such as sulphur oxides, nitrogen oxides, methane and, to a somewhat lesser extent, carbon monoxide. Also, in a confirming factor analysis, CO_2 loads on the same factor as SO_x, NO_x and CH_4. This finding leads us to the assumption that the relationship between other air pollution indicators and economic growth during the 1990s equals at least to some extent the development discovered in the analysis of the correlation between GDP and CO_2 emissions above. In the following, this analysis is extended by examining the development of air pollution in the countries of the Baltic Sea Region during the 1990s. We therefore seek to attain additional insight for the question at hand – does the economic recovery and growth of the Eastern riparian countries of the Baltic Sea necessarily lead to environmental degradation in the region?

In order to examine whether economic growth is connected to increasing air pollution, we use the concept of the decoupling factor (DF) developed by the OECD (2002).[5] Decoupling takes place if the economic growth rate is higher than the growth rate of environmental pressure in a given time period. Hence, the decoupling factor indicates the relationship between economic growth – usually indicated by GDP – and a pollution indicator. It obtains values of between −1 and 1, negative values indicating no decoupling and positive values indicating decoupling of the pollution indicator from economic growth within the given time span. Moreover, pollution may still be on the increase even if decoupling takes place; thus, the OECD distinguishes between *absolute* and *relative* decoupling. The former implies that environmental pollution decreases simultaneously to economic growth, whereas the latter means that environmental pollution still increases, although at a diminished level relative to economic growth.

The decoupling factors for all air pollutant indicators in all riparian countries are presented in Table 2.1. A brief examination reveals that all of the values are positive and thus air pollution indicators are decoupled from economic growth during the 1990s in the Baltic Sea Region.

Table 2.1 *Decoupling GDP factors from air pollution indicators, 1990–2001*

	CO_2[a]	SO_x	NO_x	CH_4	CO	VOC
Finland	0.16	0.74	0.44	0.33	0.14	0.44
Sweden	0.21	0.56	0.37	0.27	0.45	0.51
Denmark	0.31	0.89	0.45	0.18	0.38	0.41
Germany	0.25	0.90	0.55	0.51	0.66	0.63
Poland	0.39	0.67	0.58	0.56	0.68	0.53
Lithuania	0.44	0.70	0.53	0.39	0.41	0.12
Latvia	0.56	0.79	0.24	0.07	0.23	0.15
Estonia	0.44	0.61	0.40	0.51	0.56	0.59

Note: a Estonia, Latvia and Lithuania, 1992–2000; all other countries, 1990–2001.
Source: Eurostat, Maddison (2003) and author's own calculations.

The comparisons of development trends of air pollution emissions during 1990–2002 presented in Figure 2.9 show a remarkable reduction of most of the emissions in most of our cases. This is true both for the Western and the Eastern riparian countries. In the case of CO, SO_x and CH_4, Eastern countries return much higher initial levels of emissions than do Western countries, but also a more drastic reduction of the emissions during the course of the 1990s. Despite some interim variations in the declining trend, Eastern countries can be labelled as successful in reducing air pollution emissions since 1990. As in the case of CO_2 explicated earlier, we can assume the breakdown of the heavy industry to be responsible for a great part of the decreasing air pollution in Estonia, Latvia, Lithuania and Poland.

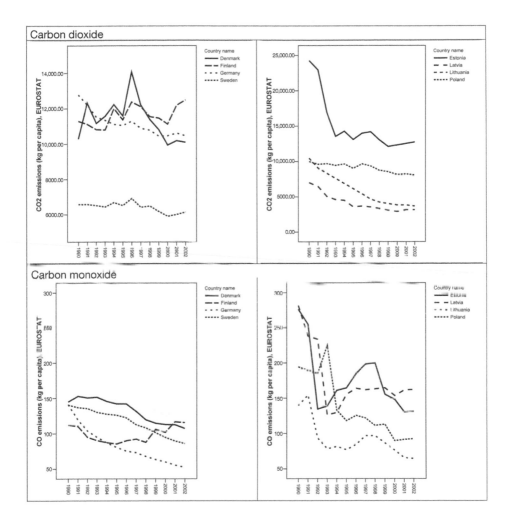

Figure 2.9 *Development of air pollution emissions, 1990–2002:*
Western riparian countries compared with Eastern riparian countries

Source: Eurostat

However, we also observe some less successful trends in NO_x and VOC among Eastern countries. Both pollutants are primarily caused by car traffic, which increased substantially in most Eastern riparian countries. Then again, new environmentally friendly technologies enabling cleaner production and fewer emissions, as well as the commitment of these countries to international environmental conventions regulating air pollution, have lead to an overall reduction in emissions. Germany, Denmark, Sweden and, with two exceptions, Finland have also managed to reduce their emissions to a remarkable extent in the course of the 1990s.

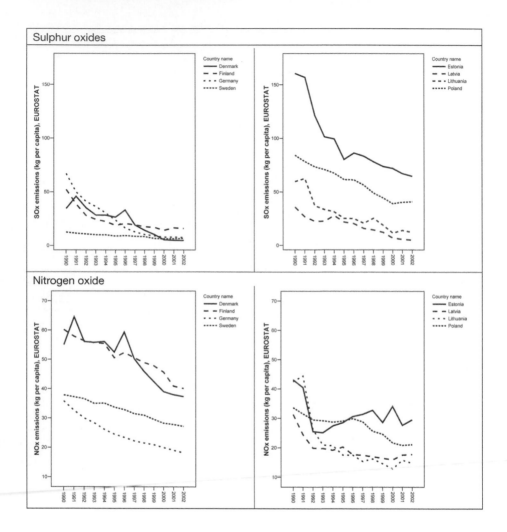

Figure 2.9 (*continued*) *Development of air pollution emissions, 1990–2002: Western riparian countries compared with Eastern riparian countries*

Source: Eurostat

The comparison between the Western and Eastern riparian countries reveals a continuing stable trend towards reduction in the Western countries. In Eastern countries, we identify a clear downward trend in air emissions as well; but fluctuation between the years is much higher. This may indicate that polluting industries and sectors are not consolidated and that there is still a question over which industrial structures and practices will prevail over the long term. Although the overall development is, without doubt, positive, there is no clear consolidation of the trends for Eastern riparian countries at the moment.

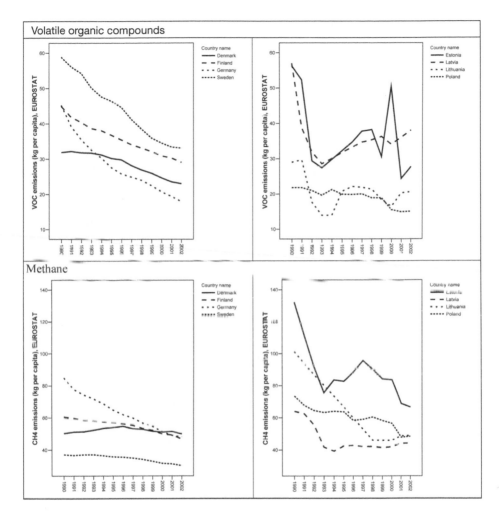

Figure 2.9 (*continued*)   *Development of air pollution emissions, 1990–2002: Western riparian countries compared with Eastern riparian countries*

Source: Eurostat

There is a clear declining and convergent trend in the development of air pollution during the 1990s. Comparison of the levels of per capita emissions in 2002 (see Figure 2.10), however, shows that major differences between the countries do exist. Interesting for our question at hand is that no clear split with regard to air pollution levels can be drawn anymore between Eastern and Western riparian countries. In addition, countries featuring negative performance concerning one air pollutant in turn perform well when another air pollutant is considered. With the exception of Estonia, which shows above average emissions levels in all air pollutants, we cannot identify univocal patterns of leaders and laggards when the overall level of air pollution is considered.

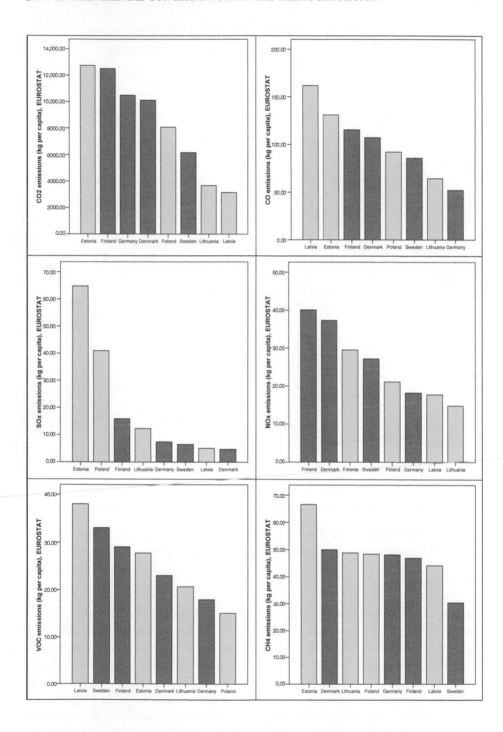

Figure 2.10 *Levels of air pollution emissions, 2002*

Source: Eurostat

This short examination of the development of air pollution and its relationship to economic growth during the 1990s seems to affirm our findings of the extended analysis of CO_2. First, economic growth is not automatically related to increasing air pollution. Second, all of the countries in the Baltic Sea Region performed well in reducing air pollution during the course of the 1990s.

CONCLUSIONS

What can we learn from comparing developments in the association between GDP and the two environmental pollution indicators – CO_2 emissions and fertilizer consumption – in countries of the Baltic Sea Region? Moreover, what does it mean for environmental governance in this region? First of all, there is a strong correlation between GDP development and CO_2 emissions, as well as between GDP development and fertilizer consumption in the Western countries analysed in this chapter. Concerning CO_2 emissions, two of the countries – Sweden and Germany – clearly follow the Kuznets curve development (i.e. the prosperity pollution development turns into prosperity clean-up after a certain level of economic welfare is reached). The two other countries – Finland and Denmark – exhibit a decoupling of GDP growth and CO_2 emissions since the 1980s, but do not confirm the prosperity clean-up hypothesis development for the time being. As for fertilizer consumption, all Western riparian countries follow a clear trend in line with the Kuznets curve development. Each country can be labelled as successful in decoupling economic growth from environmental pollution as measured by our seven indicators. However, these findings should not be viewed with too much optimism. Evidence for such a positive development regarding a larger group of Western highly industrialized countries is rather mixed (see Jahn, 2003).

Second, findings with regard to the four Eastern riparian countries of the Baltic Sea are rather encouraging when it comes to air pollution indicators. Despite rapid economic growth, these countries demonstrate steady progress in reducing CO_2, SO_x, NO_x, CH_4 and CO, as well as VOC. However, reservations remain with regard to transport emissions resulting from increased volume of car traffic. It seems, therefore, that the development of these former Eastern Bloc countries during the 1990s is not comparable with the development of Western industrialized countries in the era of industrialization and fast economic growth after World War II. This differing development may result from various factors. First, in the years following the transition, all analysed countries have gone through a drastic restructuring of their economies and a shift from the dominance of a CO_2-emitting industrial sector to a dominance of a less CO_2-intensive service sector. Second, considerable investments in new technology have been made in these countries with substantial financial and knowledge-related assistance from the Nordic countries, Germany and the EU (Turnock, 2002). Third, all countries of the Baltic Sea Region are bound by a dense net of international and regional regulatory arrangements targeting a cleaner environment. These multifaceted environmental governance tools – described in more detail elsewhere in this volume – can be expected to positively influence environmental performance. If these conclusions are also true for other countries of Central and Eastern Europe, we might expect economic growth and technological development in the region without the kind of large-scale environmental pollution as occurred in Western Europe.

However, there is a shadow over this optimistic conclusion: the prosperity cleaning-up hypothesis does not apply to the consumption of fertilizers. Rather, we observe a slight trend towards prosperity pollution, especially in Poland and Lithuania where the agricultural sector is larger than in Estonia and Latvia. However, the level of fertilizer consumption is much lower in Eastern Europe than it originally was and than it currently is in the Western Baltic Sea Region states. Nevertheless, taking the fertilizer consumption and CO_2 emissions into account leads to the conclusion that we cannot make any clear-cut statement on the relationship between economic growth and environmental pollution for the Eastern riparian countries of the Baltic Sea at the moment – in spite of strong evidence that economic development in Eastern Europe is more environmentally benign than it has been in the West over the post-World War II period.

NOTES

1   The data used in this chapter stems from two research projects funded by the German Research Foundation (DFG): Umweltbelastung als globales Phänomen (JA 638/7-1; 7-2; 7-3; 7-4) and Demokratiemuster und Leistungsfähigkeit politischer Systeme in Mittelosteuropa (JA 638/10-1; 10-2).
2   The correlation between CO_2 and SO_x, as well as between CO_2 and CH_4, but also between CO_2 and NO_x and, to some extent, CO were highly significant (e.g. at the 0.01 level) in most of our cases. The correlation between CO_2 and VOC did not show such a clear relationship.
3   For excellent data on eutrophication in the Baltic Sea Region, see, for example, the NEST database within the framework of the Maritime Research on Eutrophication (MARE) project, which is funded by the Swedish Foundation for Strategic Environmental Research (see www.mare.su.se/index.html).
4   The data for Germany during the period of 1961 to 1990 include the former German Democratic Republic.
5   The decoupling factor (DF) is defined as follows: DF = $1-[(\text{pollution indicator}_{t1}/\text{GDP}_{t1})/(\text{pollution indicator}_{t0}/\text{GDP}_{t0})]$.

REFERENCES

Andreoni, J. and Levinson, A. (1998) 'The simple analytics of the environmental Kuznets curve', National Bureau of Economic Research, Working Paper 6739, Cambridge, MA

Beckerman, W. (1992) 'Economic growth and the environment: Whose growth? Whose environment?', *World Development*, vol 20, no 4, pp481–496

Cleveland, C. J. and Ruth, M. (1998) 'Indicators of dematerialization and the materials intensity of use', *Journal of Industrial Ecology*, vol 2, no 3, pp15–50

Ehlers, P. (2001) 'Der Schutz der Ostsee – Ein Beitrag zur regionalen Zusammenarbeit', *Natur und Recht*, Heft 12, 23, pp661–666

Ekins, P. (2000) *Economic Growth and Environmental Sustainability: The Prospects for Green Growth*, Routledge, London

Grossman, G. M. and Krueger, A. B. (1995) 'Economic growth and the environment', *Quarterly Journal of Economics*, May, pp353–377

Harbaugh, W., Levinson, A. and Wilson, D. (2002) 'Re-examining the evidence for an environmental Kuznets curve', *Review of Economics and Statistics*, vol 84, no 3, pp541–551

HELCOM (2003) *The Baltic Sea Comprehensive Environmental Action Programme (JCP) – Ten Years of Implementation*, Baltic Sea Environment Proceedings, no 88, Helsinki Commission, Helsinki

Jäger, J. and O'Riordan, T. (1996) 'The history of climate change science and politics', in T. O'Riordan and J. Jäger (eds) *Politics of Climate Change: A European Perspective*, Routledge, London

Jahn, D. (1998) 'Environmental performance and policy regimes. Explaining variations in 18 OECD countries', *Policy Sciences*, vol 31, no 2, pp107–131

Jahn, D. (2001) 'Environmental and economic performance in Central Eastern Europe', Robert Schumann, Centre for Advanced Studies, European University Institute, Discussion Paper 2001/C/9·1

Jahn, D. (2003) 'Environmental pollution and economic performance in Eastern and Western Europe: Parallels and differences', Paper presented at the ECPR workshop on Explaining Environmental Policy in Central and Eastern Europe, Edinburgh, 28 March– 2 April 2003

Jahn, D. and Werz, N. (2002) 'Der Ostseeraum: Eine Zukunftsregion mit ungleichen Voraussetzungen', in D. Jahn and N. Werz (eds) *Politische Systeme und Beziehungen im Ostseeraum*, Olzog, München

Jänicke, M. and Weidner, H. (eds) (1997) *National Environmental Policies: A Comparative Study of Capacity-Building*, Julius Springer, Berlin

Jänicke, M., Mönch, H. and Binder, M. (1996) 'Umweltindikatorenprofil im Industrieländervergleich: Wohlstandsniveau und Problemstruktur', in M. Jänicke (ed) *Umweltpolitik der Industrieländer. Entwicklung – Bilanz – Erfolgsbedingungen*, Edition Sigma, Berlin

Kremser, U. and Schnug, E. (2002) 'Impact of fertilizers on aquatic ecosystems and protection of water bodies from mineral nutrients', *Landbauforschung Völkenrode (FAL Agricultural Research)*, vol 52, no 2, June, pp81–90

Kuznets, S. (1955) 'Economic growth and income inequality', *American Economic Review*, vol 445, pp1–28

Maddison, A. (2003) *The World Economy: Historical Statistics*, OECD, Paris

Mátyás, L., Kónya, L. and Macquarie, L. (1997) 'The Kuznets U-curve hypothesis: Some panel data evidence', Working Paper 7/97, Department of Econometrics and Business Statistics, Monash University, Melbourne, Australia

Milbrath, L. (1984) *Environmentalist: A Vanguard for a New Society*, State University of New York Press, Albany, NY

OECD (Organisation for Economic Co-operation and Development) (2002) *Indicators to Measure Decoupling of Environmental Pressure from Economic Growth*, OECD SG/SD(2002)1/FINAL, OECD, Paris

Paehlke, R. (1989) *Environmentalism and the Future of Progressive Politics*, Yale University Press, London

Schnaiberg, A. and Gould, K. A. (1994) *Environment and Society: The Enduring Conflict*, St Martin's, New York

Scruggs, L. (2003) *Sustaining Abundance: Environmental Performance in Industrial Democracies*, Cambridge University Press, Cambridge

Selden, T. M. and Song, D. (1994) 'Environmental quality and development: Is there a Kuznets curve for air pollution?', *Journal of Environmental Economics and Management*, vol 27, no 2, pp147–162

Stern, D. I. (2003) 'The rise and fall of the environmental Kuznets curve', Rensselaer Working Papers in Economics, no 0302, October, Rensselaer Polytechnic Institute, Troy, NY

Turnock, D. (2002) 'The central importance of the European Union', in F. W. Carter and D. Turnock (eds) *Environmental Problems of East Central Europe*, Routledge, London/New York

WCED (World Commission on Environment and Development) (1987) *Our Common Future*, Oxford University Press, Oxford

Zürn, M. (1998) *Regieren jenseits des Nationalstaates*, Suhrkamp, Frankfurt am Main

3

Institutionalization of an International Environmental Policy Regime: The Helsinki Convention, Finland and the Cold War

Tuomas Räsänen and Simo Laakkonen

INTRODUCTION

Pollution of the sea is a first-class example of the tragedy of the commons, a concept made famous by Garrett Hardin in the article of the same title published in the journal *Science* in 1968. No single state claims ownership of the open seas and, as a result, they have often been treated with contempt, as little more than convenient sinks for waste and pollution. And when the inevitable degradation occurs, states lack both the capacity and the will to claim responsibility. Since scientists discovered the degradation of the seas during the late 1960s and early 1970s, sovereign states have established dozens of international regimes for marine protection in order to overcome the tragedy of the commons.[1] Some of the agreements have been global in their nature, while others have focused on certain regional seas or restricted sea areas (Johnston, 1988, pp199–200).

By the beginning of the 1970s, the Baltic Sea was already considered one of the most polluted seas in the world. Therefore, it was not an extraordinary step for the Baltic Sea states to sign the Convention on the Protection of the Marine Environment of the Baltic Sea Area, also known as the Helsinki Convention, in 1974. However, the Helsinki Convention of 1974 differed from other regional environmental agreements in one fundamental sense. During the Cold War, the Baltic Sea was divided by the Iron Curtain, and against this background the Helsinki Convention was a remarkable achievement: at the time it was the first multilateral convention signed by members of two mutually competing military alliances. Of the seven signatories, three (the Soviet Union, Poland and the German Democratic Republic) were members of the Warsaw Pact, two (the Federal Republic of Germany and Denmark) belonged to the North Atlantic Treaty Organization (NATO), while Finland and Sweden were considered politically neutral. Moreover, the Helsinki Convention was a general text that covered all pollutants known at the time and almost all of the sea area; as such, it was the first of its kind in the world (Helsinki Convention, 1974; Ehlers, 1994, p617). The Helsinki Convention was later used as a model by other international environmental conventions, including the Convention for the Protection of the Mediterranean Sea against Pollution (Haas, 1990, p111; VanDeveer, 2002, p37).

This chapter examines the role of Finland in the process of institutionalizing an international policy regime for the protection of the Baltic Sea.[2] Our objective is to demonstrate that the Helsinki Convention of 1974 not only aimed to protect the environment, but also had a political dimension that determined the framework for the drafting and the contents of the convention. There is no doubt that the negotiations that led to the Helsinki Convention were driven by a common acceptance of a clear environmental threat; however, other more political concerns also influenced the process.

THE COLD WAR AND THE BALTIC SEA ENVIRONMENT

The relationship between the Cold War and environmental politics has been subject to a great deal of scholarly debate. Many academics have also argued that Cold War competition gradually translated into environmental cooperation in the Baltic Sea area (e.g. Rytövuori, 1980; Westing, 1989; List, 1990; Vesa, 1991; Fitzmaurice, 1992; Haas, 1993; Kohonen, 1994; VanDeveer, 2000; Darst, 2001; VanDeveer, 2002). In this chapter we will examine this complex and contested process from the point of view of the Helsinki Convention. Our study concentrates on the role of Finland in building an international institution for the Baltic Sea. During World War II, Finland was one of only three European countries involved in the conflict that were not occupied by a foreign power (the others were the Soviet Union and Great Britain). As a result of World War II, Finland had to cede land to the Soviet Union, but it remained a sovereign state with a capitalist economy and a democratic social system. On the other hand, Finland developed a unique relationship with the Soviet Union, with which Finland shared more than 1000 kilometres of land and sea border. Without doubt, close contacts between Finland and the Soviet Union benefited Finnish economic growth in the subsequent decades after World War II; however, the price Finland had to pay was the interference of the Soviet Union not only in its foreign relations, but also in its domestic policy (Jussila et al, 1999; for more critical views, see Majander, 1999, pp85–94).

The ecological condition of the Baltic Sea began to deteriorate gradually during the course of the 20th century due to urbanization, industrialization and, later, intensive agriculture (Laakkonen and Laurila, 1999; Laakkonen and Laurila, 2001). By the 1960s, some marine scientists and environmentalists argued that pollution would have destructive impacts upon ecosystems in the open sea area if the prevailing trend continued. In contrast to many other world seas, the Baltic Sea turned out to be surprisingly sensitive to pollution because it is fairly shallow, has poor water exchange with the Atlantic Ocean, and because the ice that covers the Baltic Sea during the long winter puts a further strain on the ecosystem. These factors combined to make the pollution of the Baltic Sea arguably the most serious international environmental problem in Northern Europe, directly or indirectly affecting the whole population of more than 80 million people living in the Baltic Sea basin.

Of particular interest from the viewpoint of the institutionalization of international environmental politics in the Baltic Sea Region is the hypothesis that the international political circumstances created by the Cold War played a role in the process that led to the signing of the Helsinki Convention. Several scholars have suggested that until 1972 the unsettled question over two German states' status relating to international law made the building of multilateral institutions for the Baltic Sea impossible (see Kytövuori, 1980, pp85–86; Boczek, 1981, p213; Fitzmaurice, 1992, p47; Hjorth, 1992, p164; Laakkonen, 1999, p224; VanDeveer, 2000, p14; Darst, 2001, pp55–57). As we shall see, the relationship between the Cold War and the institutionalization of international environmental regimes is far more complex. In our study, this relationship will be examined with the help of hitherto unpublished documentation and archival material. This chapter concentrates on the origins of the Helsinki Convention and particularly on an empirical examination of the views of certain actors that played an essential role in international environmental politics with respect to the convention.

The principal sources we have used in the research are the archives of the Ministry for Foreign Affairs of Finland. A significant part of this archival material consists of exchanges of information between leading Finnish politicians and civil servants of the ministry that have not been previously used in historical research. Much of the material was confidential until recently. The research material provides a fairly accurate description of Finnish viewpoints and how the political views held by other states in the Baltic Sea Region were perceived among the Finnish political elite.

In 1970, Europe celebrated European Nature Conservation Year. Concerns about the environment were also reflected in international politics, especially in the West. Environmental policy, which had previously played a marginal role in public administration, now began to be perceived as a means of responding to public pressure, but also of building up cooperation and cultivating a positive international reputation. Finland gradually began to compete against Sweden in the area of international environmental politics (Letter by Urho Kekkonen to the Ministry for Foreign Affairs, 2 March 1971, MFA, 71b, 10; Memorandum by Erkki Hedmansson, 28 May 1971, MFA, 71b, 7).

The deterioration of the state of the Baltic Sea was naturally the basic premise that created the need to develop international cooperation. But a variety of purely political factors encouraged governments to take the lead and highlighted the significance of international institutions for protecting the Baltic Sea. This was particularly true for Finland. Cooperation on environmental protection was used to foster trust between countries that belonged to opposing military alliances since environmental protection, as such, could not be deemed to involve questionable motives (Letter by Ambassador Jussi Montonen, 3 November 1972, MFA, 71b, 7; see also Lodgaard, 1989, p105; Vesa, 1991, pp75–76; Darst, 2001, pp21–24, 58–59). Consequently, the promotion of environmental cooperation was seen as useful to both the Eastern and Western factions.

In practice, the states around the Baltic Sea had engaged in certain environmental agreements even before they signed the Helsinki Convention. In 1968, Finland and the Soviet Union agreed on scientific and technological cooperation relating to the Gulf of Finland, and both parties were apparently content with this cooperation (Records of the negotiations for the establishment of cooperation on marine research in the Gulf of Finland, 2 April 1968, FIMR, Gulf of Finland Working Group, 1; Räsänen and Laakkonen, 2007a). Some years later, Sweden and the Soviet Union engaged in similar research cooperation pertaining to the Baltic Sea. Finland and Sweden had also traditionally worked together extensively on environmental issues, although the two countries did not sign an actual agreement on cooperation regarding the Baltic Sea environment until 1972. The Soviet Union, the German Democratic Republic (GDR) and Poland had signed the declaration concerning the use of the continental shelf, while Sweden and Denmark had cooperated on the protection of the sound of Öresund since 1960 (Conference Plan for the protection of the Baltic Sea: background information, 17 August 1972, MFA, 71b, 10).[3] All of these agreements provided a valuable point of departure for building a more comprehensive environmental regime in the Baltic Sea area; but, as seen from the list above, none of the existing agreements reached beyond the boundaries of the military alliances. Yet, for environmental actors it was clear that cooperation between all states was necessary in order to curb emissions that degraded Baltic Sea ecosystems.

FINLAND, SWEDEN AND THE GERMAN QUESTION

One crucial result from the political division of the Baltic Sea was that the two neutral countries in the region, Finland and Sweden, were in the best position to take the lead in intergovernmental negotiations on cooperation related to the area without provoking political conflicts between the superpowers. Sweden took the first initiative. It advocated an agreement between the states around the Baltic Sea to protect it from oil discharges from ships. However, two meetings held in 1969 and 1970 in Visby, Sweden, resulted only in two separate protocols that stressed the importance of joint efforts in matters relating to the environmental protection of the Baltic Sea and which were hoped to serve as a basis for possible future agreement. No international agreement was signed because the NATO countries in the region felt unable to endorse an agreement with the GDR (Memorandum by Arvo Harjula and Arvo Karjalainen, 24 September 1969, MFA, 71b, 26; Travel reports by Arvo Harjula, 6 October 1969 and 6 October 1970, MFA, 71b, 26).

The underlying issue was the division of Germany. As the 1970s began, neither the GDR nor the Federal Republic of Germany (FRG) officially recognized each other. Their respective allies also failed to recognize the sovereignty of the opposing states. This deadlock, known as the German question, thwarted the signing of any multilateral agreements on the protection of the Baltic Sea, for the signing of governmental treaty between all the Baltic Sea states would have meant mutual recognition between the two Germanys. The Soviet Union set the participation of GDR as a condition for signing any convention on international environmental protection (Memorandum by Paul Gustafsson, 2 April 1969, MFA, 71b, 26).

The attitudes of the Finnish and Swedish governments differed on the German question. Sweden, along with other Western countries, had officially recognized the FRG, but not the GDR, shortly after its creation. Finland, in contrast, had not recognized either state, so the Finnish government was able to promote multilateral agreements on environmental protection without the disadvantage faced by the Swedes. On the other hand, both the West and the East had some reservations about Finnish neutrality. The West saw Finland as being firmly in the Soviet sphere of influence, which raised doubts about the ultimate aims of Finland. The Eastern faction felt that things were as they should be; all they asked Finland to do was to stop emphasizing its neutral status (Suomi, 1996, p307). Indeed, the key difference between Finland and Sweden was Finland's good relations with the Soviet Union. Since Sweden appeared to the Eastern Bloc as being biased on the German question, the initiative in establishing an international institution for the environmental protection of the Baltic Sea shifted from the Swedish government to the Finns.

Soon after the Visby meetings, officials in the Ministry for Foreign Affairs of Finland began to generate ideas for a more comprehensive convention on the protection of the Baltic Sea. The Finns believed that the Soviet Union would look more favourably on initiatives taken by Finland than those taken by Sweden (Letter by Urho Kekkonen to the Ministry for Foreign Affairs, 2 March 1971, MFA, 71b, 10). After all, it was common knowledge after the Visby meetings that Sweden would not sign an intergovernmental agreement in which GDR was a contracting party, whereas Finland was prepared to enter into 'any multilateral agreement, if [other] countries are also capable of entering into it' (Memorandum by Risto Hyvärinen, 29 March 1971, MFA, 71b, 7). For as long as the German question caused friction between the Baltic coastal states, Finland remained the most acceptable choice to convene an intergovernmental conference on the protection of the Baltic Sea.

The need to cooperate on environmental issues thus enabled Finland to pose as peace-maker between the states around the Baltic Sea and to strengthen its policy of neutrality and its role as an active international player. The Finns hoped that environmental cooperation would later have a positive impact upon the organization of the Conference on Security and Cooperation in Europe (CSCE), in which both Eastern and Western European countries were expected to participate. Initially, at the proposal of the Soviet Union, Finland began to plan the organization of the CSCE in 1969. Finland's highest leaders rated the organization of the CSCE, along with the nurturing of relations with the Soviet Union, as the most important international goal of Finnish foreign policy at the time.

THE FINNISH PROPOSAL FOR TRILATERAL AGREEMENT

At the beginning of 1971, the Finnish government believed that preparations for the Convention on the Protection of the Marine Environment of the Baltic Sea Area should begin immediately because it seemed that the international political obstacles could be overcome (Memorandum by Kurt Uggeldahl, 7 July 1971, MFA, 71b, 10).[4] The German question, in particular, was heading for a resolution: in January 1971, the leaders of the GDR and FRG, Willi Stoph and Willy Brandt, had initiated negotiations on the normalization of relations between the two countries. Significant progress had also been made in the border dispute between Germany and Poland, which had led to a formal recognition of the Oder-Neisse line by FRG in December 1970 (Glees, 1996, pp172–175; Hentilä, 2003, pp114–116).

The idea of an international conference on the pollution of the Baltic Sea was realized in the spring of 1971. During a visit to Moscow, Finnish President Urho Kekkonen expressed his country's willingness to host such a conference, and the Soviet Prime Minister Alexei Kosygin gave the green light to the idea (Letter by Urho Kekkonen to the Ministry for Foreign Affairs, 2 March 1971, MFA, 71b, 10). Finland wanted to take a cautious approach and started by sounding out the views of Sweden and Denmark, which were considered politically less sensitive than other countries in the region. In consultations that took place in July 1971, both Sweden and Denmark reiterated their previous position: the initiative was welcome, in principle; but the GDR could not be a party to an intergovernmental convention. However, the consultations concluded with an agreement that Finland would extend a general conference invitation to all Baltic coastal states. Preparations for the conference would take time, and the German question would perhaps be resolved, in the meantime, in a manner that would be satisfactory to everyone involved (Memorandum by Kurt Uggeldahl, 7 July 1971, MFA, 71b, 10). An un-official invitation was immediately issued in the summer of 1971 after the meeting of the representatives of Finland, Sweden and Denmark (Memorandum of 17 August 1971, MFA, 71b, 10). Later in the same autumn the matter was made public in Ottawa during preparatory consultations for the first United Nations Conference on the Human Environment. The invitation was also repeated in Stockholm in 1972 during the actual conference (Memorandum of 17 August 1971, MFA, 71b, 10; Memorandum by Henrik Blomstedt, 15 December 1972, MFA, 71b, 10).

At the same time, pressure to promote environmental protection was growing, especially after the media became aware of plans for the Helsinki conference.[5] Moreover, alarming headlines on the deterioration of the marine environment circulated in the press: 'Life beginning to fade in the Baltic Sea' (*Turun Sanomat*, 11 June 1971). The media sent a clear message to the decision-makers: the states around the Baltic Sea had to work together against pollution, or soon they would have nothing to save (see, for example, *Turun Sanomat*, 20 September 1970; *Turun Sanomat*, 11 June 1971; *Turun Sanomat*, 11 July 1971; *Uusi Suomi*, 15 September 1970; *Uusi Suomi*, 7 July 1971).

The Legal Department of the Ministry for Foreign Affairs of Finland was looking for alternatives to circumvent the German problem. One of the ideas was to enter into a non-governmental agreement between expert organizations. Another idea was put forward to sign several multilateral agreements that would have the same content. These proposals, among others, were rejected; but one alternative was developed further. This involved signing a trilateral agreement between Sweden and Finland, and the Soviet Union (Memorandum by Kurt Uggeldahl, 7 July 1971, MFA, 71b, 10).

The idea of trilateral cooperation between Finland, Sweden and the Soviet Union was initially suggested by marine researchers. Bilateral research cooperation between Finland and the Soviet Union had functioned well. This led in 1970 to discussions on extending the cooperation to include a third state. The Ministry for Foreign Affairs also began to chart opportunities for cooperation between Sweden, Finland and the Soviet Union (Letter by Aarno Voipio and Ilmo Hela to the Ministry of Trade and Industry, 1 December 1970, CS/MTI, 63/570/70; Letter from the Ministry of Trade and Industry to the Ministry for Foreign Affairs, 7 December 1971, MFA, 71b, 7; Letter by Aarno Voipio to the Ministry of Trade and Industry, 10 September 1971, CS/MTI, 57/570/1971). The coastal area belonging to Sweden, Finland and the Soviet Union covered four-fifths of the Baltic coastal line, and at the time it was thought that most of the pollution deriving from human activities and ending up in the Baltic Sea came from these states. The resolution of environmental problems could be addressed through trilateral cooperation until political conflicts had been settled and a more comprehensive agreement could be reached. It was hoped that trilateral cooperation would encourage other states to gradually join cooperation on marine protection, which would lead to a convention on the protection of the marine environment of the Baltic Sea, signed by all Baltic coastal states (Memorandum by Henrik Blomstedt, 24 November 1971, MFA, 71b, 7; Explanatory memorandum by Henrik Blomstedt, 27 December 1971, MFA, 71b, 7).

The project outlined by the Finnish government received a positive initial response from the Soviet Union and Sweden. However, the Soviet representatives stressed in no uncertain terms that if the Finnish proposal included the idea of an intergovernmental trilateral treaty, the Soviet Union would reject the initiative. In contrast, the Soviets could accept practical research cooperation without governmental agreement; but even this arrangement should not involve any obstacles for all other Baltic States to sign on at a later date (Memorandum by Henrik Blomstedt, 24 November 1971, MFA, 71b, 7; Notes by Henrik Blomstedt of a telephone discussion with Myrsten, Swedish Counsellor for Foreign Affairs, 15 December 1971, MFA, 71b, 7).

Building on the initial discussions, Finland hoped to make rapid progress; but the project met with resistance. When Åke Wihtol, the deputy director-general of the Ministry for Foreign Affairs, put forward the above-mentioned proposal for cooperation to the Soviet ambassador in Helsinki, he was met with a cool response. The ambassador deemed the Finnish proposal to be an attempt to circumvent political realities and accused the Finns of wanting to initiate negotiations without the participation of all the Baltic coastal states (Memorandum by Åke Wihtol, 22 December 1971, MFA, 71b, 7). Despite the reluctance of the Soviet ambassador, Finland decided to submit an initiative to the Soviet Union on trilateral cooperation. In April 1972, the Soviet Union issued its response: the participation of the GDR was a necessary pre-condition for the resolution of any international affairs relating to the protection of the Baltic Sea environment (Translation into Finnish of the Soviet response to the Finnish initiative on trilateral cooperation, 7 April 1972, MFA, 71b, 7).

The Soviet environmental policy concerning the Baltic Sea was tied in with the German question. Since 1969, the Soviet Union had tried to persuade Finland to recognize the GDR, hoping that other Western countries would follow suit (Suomi, 1996, p307). It was hoped that a multilateral agreement on the protection of the Baltic Sea would promote a resolution to the GDR question. Finland was trying to strike a balance in these conflicting circumstances. According to the consensus in Finland, 'the matter at stake is the degradation of the Baltic Sea, not the German question' (Memorandum by Kurt Uggeldahl, 8 March 1971, MFA, 71b, 7). But the German question was nevertheless part of the international political realities of the Baltic Sea area, and it could not be ignored. The Soviet Union demanded that Finland unilaterally recognize the GDR, whereas the Finns feared that such a decision might bring about retaliation from the Western powers. The FRG let it be understood that if Finland were to recognize the GDR under pressure from the Soviet Union, this would not only jeopardize Finland's position as the designated host of the Conference on Security and Cooperation in Europe, but would also compromise Finland as a host of any subsequent multilateral conference (Suomi, 1998, pp209-210).[6] The Finnish government refused to yield to Soviet demands.

At the beginning of 1972, some people continued to hold out hope that the precarious state of the Baltic Sea environment would be discussed in Stockholm at the first UN Conference on the Human Environment in June 1972. However, shortly before the conference was due to begin the Eastern Bloc countries stated that because the GDR was not allowed to participate in the conference, they would not be sending their representatives to Stockholm (Hjorth, 1992, p17). As the Soviet Union and Poland boycotted the Stockholm conference, it was impossible to address the Baltic Sea issue effectively. Obviously, the failure of the Stockholm conference added weight to the Finnish initiative.

Finnish patience was rewarded when the FRG reported, at the end of October 1972, that a resolution to the German question was close at hand. The final break-through came quickly afterwards: on 7 November 1972, the representatives of the FRG and GDR announced that they had reached agreement on fundamental issues. The two countries signed the agreement known as the Basic Treaty on 21 December 1972 in which the FRG finally recognized the GDR as an independent state (Hentilä, 2002, p278). The resolution of the German question removed the most difficult obstacle to international cooperation on the protection of the Baltic Sea.

FINLAND'S CHANCES OF HOSTING THE CONFERENCE ARE STILL UNCERTAIN

Once the political deadlock seemed to be broken, Finland immediately sent an official conference invitation via its embassies to all the states around the Baltic Sea. The Finnish government hurried preparations to host the conference as they feared that the settlement of the German question, and the resulting recognition of the GDR by other Nordic nations, would compromise the Finnish political edge. With the loss of this political edge it was thought that Sweden, with its greater resources, would be a more suitable conference venue. There were also rumours that in bilateral consultations with Sweden, Poland had expressed its interest in arranging such a conference (Copy of a secret telegraph, Finnish Embassy in Warsaw, 24 January 1973, MFA, 71b, 11).

The fear of losing the conference to Sweden or some other state proved to be unwarranted as Finland's initiative gradually started advancing in the Baltic Sea states shortly after the treaty between the FRG and GDR had been signed. The first ones to announce their willingness to participate in a conference hosted by Finland were Sweden, Denmark and the GDR (Memorandum by Henrik Blomstedt, 30 January 1973, MFA, 71b, 11). As for the FRG, the Soviet Union and Poland, Finnish officials conceived that they clearly waited for each other's responses to the conference invitation. Poland waited for the Soviet Union's acceptance, the Soviet Union waited for the FRG's acceptance and the FRG was still hesitant about co-signing an agreement with the GDR, despite the basic treaty that had been concluded between the two countries (Memorandum by Gylling, 27 February 1973, MFA, 71b, 11).

Even though such uncertainty prevailed and the responses of some coastal states were delayed, Finland started drafting the text of the convention since it had agreed with Sweden, Denmark and the Soviet Union that Finnish experts would form a Baltic Sea commission to draw up a draft of the convention text (Henrik Blomstedt's letters to the Finnish Embassies around the Baltic Sea, MFA, 71b, 11). At the same time, it was agreed to arrange an expert meeting in late May and early June 1973 (Blomstedt: copy of a secret telegraph, 2 March 1973, MFA, 71b, 10).

In the course of spring 1973, those countries that had previously been hesitant about the conference gave their acceptance to the arrangement of the expert meeting. Poland's response came in early March, the FRG responded in late March and the Soviet Union's final response came as late as 15 May 1973, only two weeks before the expert meeting was scheduled to take place (Copy of a letter sent as secret telegraph by the Finnish Embassy in Warsaw, 3 March 1973, MFA, 71b, 11; Letter by the Finnish Embassy in Bonn to the MFA, 21 March 1973, MFA, 71b, 11; Memorandum by Paul Gustafsson, 15 May 1973, MFA, 71b, 11). Thus, the contracting parties were finally known in summer 1973.

DRAFTING THE CONVENTION

Shortly before the expert meeting, alarming rumours reached Finland. In contacts between marine researchers, Finns had been led to understand that the socialist countries were willing to promote bilateral environmental protection, but that multilateral agreements would be out of the question. Also, the socialist countries opposed the idea that the convention would include information on pollutants discharged into the sea, suggesting that this would be tantamount to industrial espionage. Consequently, they organized that the convention should concentrate on airborne pollutants and the pollution drifting into the Baltic Sea from the North Sea (Ilmo Hela's letter to Paul Gustafsson, 16 May 1973, MFA, 71b, 11).[7] While the preparatory Baltic Sea commission had suggested that all known sources of pollution should be included in the convention, Germany, for its part, suggested that the convention be limited to the prevention of oil pollution (Memorandum by the Baltic Sea Commission, 12 February 1973, MFA, 71b, 11; Henrik Blomsted's old boy correspondence to I. M. Behnke, Denmark's ambassador to Helsinki, 29 March 1973, MFA, 71b, 11; Notes from the meeting of the MFA Legal Department, 30 March 1973, MFA, 71b, 11).

Finland decided to go ahead with the preparation of a more comprehensive text of the convention and the expert meeting was held, as planned, in Helsinki between 28 May and 2 June 1973. The differences and conflicts that had surfaced earlier were also discussed at the meeting. Finland's representative Paul Gustafsson thought that the FRG's reluctant attitude towards the conference became clearly evident from the conduct of its representatives. Gustafsson suspected, however, that the representatives had been forbidden to thwart the plans for the conference if the other coastal states were favourably disposed to the idea. The Soviet Union favoured the organization of the conference and emphasized the importance of the convention to the country. On the other hand, the Soviet Union also insisted that the sea area within 12 nautical miles from respective coastlines should not be included in the geographical coverage of the convention; instead, each country would be independently responsible for monitoring and protective action in its internal waters (Notes by Holger Rotkirch of the expert meeting, 28 May 1973, MFA, 71b, 11; Memorandum by Paul Gustafsson, 5 June 1973, MFA, 71b, 11).

Finland offered, once again, to host the Diplomatic Conference on the Protection of the Marine Environment of the Baltic Sea Area, and the participants agreed that the conference would be held in early 1974. At the same time, a preparatory committee was established and its first meeting was set for November 1973. Finland agreed to prepare a draft of the convention text on the basis of suggestions from all Baltic coastal states (Memorandum, 4 June 1973, MFA, 71b, 11; Memorandum by Gustafsson, 11 June 1973, MFA, 71b, 10).

The draft was finished by the beginning of October and when the representatives of the coastal states met during the conference of the International Maritime Organization, held in London, the comments were almost unanimously positive (Communication of negotiations between the Baltic Sea states on 8 October 1973, MFA, 71b, 12). The only controversial issue that remained was the inclusion of territorial waters in the geographical coverage of the convention. In bilateral consultations, Finland and Sweden had agreed that the inclusion of territorial waters was of the utmost importance in the prevention of pollution (Memorandum by Unto Turunen, 28 June 1973, MFA, 71b, 11). The Soviet Union, for its part, held its ground and insisted that territorial waters were not to be included in the convention (Soviet Union's response to the draft of the convention prepared by Finland, 3 November 1973, MFA, 71b, 12). In the end, the Soviet Union had its way and the security considerations of a superpower prevailed over intentions for more transparent coastal activities.

The text of the Convention on the Protection of the Marine Environment of the Baltic Sea Area was almost finished in the autumn of 1973. In the consultations between the coastal states held in November 1973, it became evident that the focus of discussions on the convention had shifted from the problems of international politics to issues of environmental protection. Since the most pressing political controversies were fortunately left behind and the consensus was attained vis-à-vis environmental questions, the text was handed over to the legal authorities for final polishing, and by February 1974 the guidelines approved by all states had been formulated into a single document. The meeting of the legal authorities finally dared to announce the long-awaited news: 'each participant state will sign the convention after the conference' (Memorandum by Gustafsson, 15 November 1973, MFA, 71b, 12; Memorandum by Pertti Harvola, 28 February 1974, MFA, 71b, 6).

IMPLEMENTING THE HELSINKI CONVENTION

The long and complex process leading to the Helsinki Convention was brought to a close in 1974. At the end of the conference held in Helsinki between 18 and 22 March 1974, Finnish Foreign Trade Minister Jermu Laine was able to propose to the representatives of the Baltic Sea states that the international Convention on the Protection of the Marine Environment of the Baltic Sea Area be approved. All the states concerned were unanimous in approving the convention (Address by minister Jermu Laine, 22 March 1974, MFA, 71b, 6). After careful preparation, the convention was finalized and the signatories could join forces in seeking solutions to the environmental problems plaguing the Baltic Sea. The principle of building trust between states was noted in the convention: 'Awareness of the significance of intergovernmental cooperation in the protection of the marine environment of the Baltic Sea as an integral part of the peaceful cooperation and mutual understanding between the nations of Europe' (Helsinki Convention, 1974).

But the convention was only a prelude to practical cooperation. In practice, international cooperation in marine protection required assessing and harmonizing the scientific and technological practices of the signatory countries, which proved to be a demanding task in such a divided area, both politically and culturally. For example, debates dragged on for years over the terms of reference of the convention, such as the definition of pollution (Darst, 2001, p63).

Consequently, the Helsinki conference established the Baltic Marine Environment Protection Commission (Helsinki Commission/HELCOM), which was to start operations once all of the seven signatories had ratified the convention. While the ratification was in process, the commission was to be substituted by an interim protection commission, which was assigned with the task of establishing common procedures, practices and criteria for the launching of concrete protective actions. By 1980, all the coastal states had ratified the convention – thus, the Convention on the Protection of the Marine Environment of the Baltic Sea Area officially entered into force.

The delicate background of international politics behind the Helsinki Convention was, however, reflected in restrictions of authority. Above all, the decisions of the Helsinki Commission were only non-binding recommendations. The Helsinki Commission responsible for implementing the convention did not turn into a supranational organization with the power to compel the contracting parties to engage in protective actions. Moreover, questions of responsibility and compensation in case of accidents were not adequately taken into account when the convention text was drafted. From a regional perspective, the convention was restricted most by the Soviet demand, based on the national security doctrine of the Soviet Union, not to include territorial waters in the contents of the convention. As a consequence, the authority of the Helsinki Commission did not extend to the coastal zone, which was the most polluted area.

Because of the lack of instruments to force the states to act in accordance with the letter of the convention, environmental policy regime remained deficient for the full duration of the Cold War. None of the contracting parties were able to fulfil all of the HELCOM recommendations. Yet, there were huge differences between individual states. Some of the countries were active in adding the recommendations into their domestic legislation and creating new means to combat pollution, while others were little more than bystanders (Hjorth, 1992). Some countries almost certainly violated fundamental provisions of the convention by, for example, continuing secretly to emit banned toxic substances such as DDT and PCB (e.g. Fedorov, 1997; VanDeveer, 2002, p33). When the political situation in the Baltic Sea Region yet again changed with the fall of the Soviet Union, the process of democratization within the former Communist Bloc and the re-independence of the Baltic states, the Helsinki Convention was revised and re-ratified in 1992.

CONCLUSIONS

In the spring of 1974, the states bordering the Baltic Sea could celebrate a joint success. Finland had a particular cause for celebration: after all, the convention was the result of years of hard work by Finnish politicians and officials. In addition, Finland proved able to overcome Sweden in the competition for international environmental agreements. The significant achievement was, however, that with the Helsinki Convention and under the pretext of environmental protection, cooperation between the Baltic Sea states rose to a new level, for the convention was the first multilateral agreement between members of two competing military alliances.

The convention was drafted against the backdrop of Cold War rivalry in the Baltic. The Soviet Union paid lip service to environmental concerns with a view to advancing longstanding political goals – notably, the desire to achieve political recognition for the GDR. Thus, it may be argued that the Soviet Union used environmental politics as a new tool of power politics.

Sweden had initiated efforts to work out an international agreement for the protection of the Baltic Sea, but had failed since it would not recognize the GDR. The first United Nations Conference on the Human Environment, hosted by Sweden in 1972, failed partly for the same reason: the Soviet Union and the other Eastern Bloc countries boycotted the conference. Therefore, the initiative for a convention on the protection of the Baltic Sea shifted from Sweden to Finland, which in the international political playoffs remained the only country in the Baltic Sea area credible enough to negotiate with both the East and the West.

The promotion of the protection of the Baltic Sea also suited the agenda of the political leadership of Finland, which realized that environmental issues could be used to further the country's main diplomatic objective, which was hosting the Conference on Security and Cooperation in Europe. The building of an international institution for environmental protection brought the Eastern and Western Blocs closer to each other and thus paved the way for a greater political end: the stabilization of the political situation in all of Northern Europe.

The Cold War and international environmental politics crossed paths in exceptionally favourable circumstances at the beginning of the 1970s, for at the time, a process of détente was well under way in the Baltic Sea Region. In hindsight, it was fortunate that the negotiations that led to the institutionalization of environmental policy regime in the Baltic Sea Region took place at the same time. For the first time in international relations, the climate of détente culminated in concrete environmental cooperation over the Iron Curtain. However, the mid 1970s could be characterized as a watershed of détente in Europe. After that, it was clear that both the spirit and momentum of détente were beginning to falter (Nelson, 1995, p147; Hanhimäki, 1998, pp326–327).

We conclude that the relation between the Cold War and the environment was twofold. On the one hand, the Cold War thwarted environmental cooperation as far as the German question remained unsettled. On the other hand, without political division in the Baltic Sea Region, it would have been questionable whether Finland would ever have taken such an active role in promoting international institutionalization for marine protection. The Helsinki Convention of 1974 was diluted by certain fundamental deficiencies, and until the end of the Cold War its strength was restricted by clear indifference towards the work of the Helsinki Commission on the part of some participant states. Nonetheless, the convention established a vital foundation for an international environmental regime, without which the implementation of the revised Helsinki Convention of 1992 would have been much more problematic.

NOTES

1   The first international agreement for the protection of the marine environment was already signed already in 1954 (the International Convention for the Prevention of Pollution of the Sea by Oil, or OILPOL); but the real growth in the number of agreements began only after the late 1960s.
2   For previous discussion on this issue see Räsänen and Laakkonen, 2007b.
3   In addition, there were various agreements between the Baltic Sea countries concerning fishery and the protection of fish stocks.
4   According to Juhani Suomi (1996, pp580–586), President Kekkonen predicted, as early as autumn 1970, that the German question would soon be settled. Over the following year, the Ministry for Foreign Affairs also understood that policies related to the recognition of the GDR and the FRG could be revised.

5   The role of the media in promoting environmental protection should not be underestimated. The Finnish representatives, among others, in the Visby meetings considered the Swedish initiatives, which were doomed to failure from the start, primarily as the reactions of politicians to increased pressure, not as genuine attempts to resolve environmental problems or further develop the international agreement system (see Arvo Harjula's travel report, 6 October 1970, MFA, 71b, 26).
6   The warning issued by the FRG concerned the so-called German package, in which Finland would have recognized both the FRG and the GDR.
7   For Finland and Sweden, the exchange of information concerning pollutant loads was one of the primary purposes of the proposed international organization.

REFERENCES

Boczek, B. (1981) 'The Baltic Sea: A study in marine regionalism', *German Yearbook of International Law*, vol 23, pp196–230

Darst, R. G. (2001) *Smokestack Diplomacy: Cooperation and Conflict in East-West Environmental Politics*, MIT Press, Cambridge, MA

Ehlers, P. (1994), 'The Baltic Sea area: Convention on the protection of the Baltic Sea area (Helsinki Convention) of 1974 and the revised convention of 1992', *Marine Pollution Bulletin*, vol 29, no 6–12, pp617–621

Fedorov, L. A. (1997) 'Officially banned, unofficially used', *Global Pesticide Campaigner*, vol 7, no 4, www.panna.org/resources/pestis/PESTIS.1997.103.html, accessed 10 September 2007

Fitzmaurice, M. (1992) *International Legal Problems of the Environmental Protection of the Baltic Sea*, Kluwer Law International, Dordrecht, The Netherlands, Boston, MA, and London, UK

Glees, A. (1996) *Reinventing Germany. German Political Development since 1945*, Berg Publishers, Oxford, UK

Haas, P. M. (1990) *Saving the Mediterranean: The Politics of International Environmental Cooperation*, Columbia University Press, New York, NY

Haas, P. M. (1993) 'Protecting the Baltic and North Seas', in P. M. Haas, R. O. Keohane, M. A. Levy (eds) *Institutions for the Earth: Sources of Effective International Environmental Protection*, MIT Press, Cambridge, MA

Hanhimäki, J. M. (1995) 'Ironies and turning points: Détente in perspective', in O. A. Westad (ed) *Reviewing the Cold War: Approaches, Interpretations, Theory*, Cass, London, UK

Hardin, G. (1968) 'The tragedy of the commons', *Science*, vol 162, pp1243–1248

Helsinki Convention (1974) *Final Act of the Diplomatic Conference on the Protection of the Marine Environment of the Baltic Sea Area*, Helsinki, 22 March 1974

Hentilä, S. (2002) 'Willy Brandtin uusi idänpolitiikka: DDR ja Suomi', in E. Sundbäck (ed) *'Muille maille vierahille...': Kalervo Hovi ja yleinen historia*, Turku Historical Society , Turku, Finland

Hentilä, S. (2003) *Kaksi Saksaa ja Suomi. Saksan kysymys Suomen puolueettomuuspolitiikan haasteena*, SKS, Helsinki, Finland

Hjorth, R. (1992) *Building International Institutions for Environmental Protection: The Case of Baltic Sea Environmental Cooperation*, Kanaltryckeriet, Motala, Sweden

Johnston, D. M. (1988) 'Marine pollution agreements: Successes and problems', in J. E. Carroll (ed) *International Environmental Diplomacy: The Management and Resolution of Transfrontier Environmental Problems*, Cambridge University Press, Cambridge, UK

Jussila, O., Hentilä, S. and Nevakivi, J. (1999) *From Grand Duchy to a Modern State: A Political History of Finland since 1809*, Hurst and Company, London, UK

Kohonen, T. (1994) 'Regional environmental policies in Europe: Baltic/Nordic cooperation', in O. Höll (ed) *Environmental Cooperation in Europe: The Political Dimension*, Westview Press, Boulder, CO

Laakkonen, S. (1999) 'Harmaat aallot. Ympäristönsuojelun tulo Suomeen', in S. Laakkonen, S. Laurila and M. Rahikainen (eds) *Harmaat aallot. Ympäristönsuojelun tulo Suomeen*, Finnish Historical Society, Vammala, Finland

Laakkonen, S. and Laurila, S. (eds) (1999) 'The history of urban water management in the Baltic Sea region', *European Water Management*, vol 2, no 4, pp29–76

Laakkonen, S. and Laurila, S. (eds) (2001) 'Man and the Baltic Sea', *Ambio – A Journal of the Human Environment*, vol 30, no 4–5, pp263–326

List, M. (1990) 'Cleaning up the Baltic: A case study in East–West environmental co-operation', in V. Rittberger (ed) *International Regimes in East–West Politics*, Pinter Publishers, London, UK, and New York, NY

Lodgaard, S. (1989) 'Confidence building in the Baltic region', in A. H. Westing (ed) *Comprehensive Security for the Baltic: An Environmental Approach*, Sage, London, UK

Majander, M. (1999) *The Paradoxes of Finlandization, in Northern Dimensions*, The Finnish Institute of International Affairs, Helsinki, Finland

Nelson, K. (1995) *The Making of Détente: Soviet-American Relations in the Shadow of Vietnam*, Johns Hopkins University Press, Baltimore, MD, and London, UK

Räsänen, T. and Laakkonen, S. (2007a) 'Suomen ja Neuvostoliiton ympäristöyhteistyön alkuvaiheet', *Historiallinen Aikakauskirja*, vol 105, no 1, pp43–56

Räsänen, T. and Laakkonen, S. (2007b) 'Cold war and the environment: The role of Finland in international environmental politics in the Baltic Sea Region', *Ambio*, vol 36, nos 2–3, pp229–236

Rytövuori, H. (1980) 'Structures of détente and ecological interdependence: Cooperation in the Baltic Sea area for the protection of marine environment and living resources', *Cooperation and Conflict: Nordic Journal of International Politics*, vol 15, pp85–102

Suomi, J. (1996) *Taistelu puolueettomuudesta: Urho Kekkonen 1968–1972*, Otava, Helsinki, Finland

Suomi, J. (1998) *Liennytyksen akanvirrassa: Urho Kekkonen 1972–1976*, Otava, Helsinki, Finland

VanDeveer, S. D. (2000) 'Protecting Europe's seas', *Environment*, vol 42, no 6, pp10–26

VanDeveer, S. D. (2002), 'Environmental cooperation and regional peace: Baltic politics, programs, and prospects', in K. Conca and G. D. Dabelko (eds) *Environmental Peacemaking*, Johns Hopkins University Press, Washington, DC

Vesa, U. (1991) 'Environmental security and the Baltic Sea region', in P. Joenniemi (ed) *Co-Operation in the Baltic Sea Region; Needs and Prospects*, Tampere Peace Research Institute, Tampere, Finland

Westing, A. H. (ed) (1989) *Comprehensive Security for the Baltic: An Environmental Approach*, Sage, London, UK

Archival sources

MFA: Archives of the Ministry for Foreign Affairs, Helsinki, Finland
FIMR: Archives of the Finnish Institute of Marine Research, Helsinki, Finland
CS/MTI: Archives of the Council of State, the Ministry of Trade and Industry, Helsinki, Finland

Newspapers

Turun Sanomat, Turku, Finland
Uusi Suomi, Helsinki, Finland

4

Environmental Concerns within the Baltic Sea Region: A Nordic–Baltic Comparison

Ann-Sofie Hermanson

INTRODUCTION

The overriding question for this chapter – whether early institutionalization and a slow development of environmental institutions create outputs different from those of systems marked by a rapid transition – is not easily answered. This is also what Jänicke (1996) states; it is difficult to test the hypothesis that early institutionalization leads to a better environmental policy record. However, the efforts of Jänicke (1997, 2002) guide this attempt to find out whether there is a north-western and a south-eastern tradition of environmental politics in the Baltic Sea Region. The first part of this chapter concentrates on the political systems and the environment of the region. In the second part, the focus is on the region's citizens and their environmental concerns.

The Nordic countries and Germany are known to be forerunners in environmental matters (Andersen and Liefferink, 1997). However, many countries developed environmental administrative units around the Stockholm Conference held in 1972. East European countries (e.g. the GDR and Poland) did so too during the early 1970s, but very unsuccessfully in many respects (Andersson, 2002). 'This suggests a differentiation between formal and substantive institutionalisation. In democratic systems it may be easier to create institutions in the strict sense, resulting in stable, calculable, internalised and valued patterns of behaviour', says Jänicke (1997, p10), referring to Huntington (1965). The difference stems from whether environmental institutions have stringently protected jurisdiction, support from an environmental/ecology movement, and an independent media or not (Jänicke, 1997, p10).

This chapter focuses on traditional nation state-centred capacities for the environment, constituted by the strength of governmental and non-governmental proponents to conduct environmental protection in the Baltic Sea Region. The media is not studied due to the scope of the research. In analysing structural framework conditions, it is absolutely necessary to bear in mind the different meanings of certain concepts in different contexts. Nevertheless, this chapter argues that comparative research following Jänicke's list of institutionalization steps brings understanding of the similarities and differences in this field. A description of concepts and phases of development clarifies the pinpointed events. To begin with, the chapter, therefore, has a look at different institutionalization steps in terms of environmental proponents, identifying them and their activities and gains in the Baltic Sea Region. Special emphasis is laid on the Nordic countries, while Chapter 5 focuses on the development of environmental governance in the Baltic states.[1] The chapter then examines public environmental concern in the Baltic Sea Region.

Environmental actors, both government institutions and movements (non-governmental proponents in terms of Jänicke's definition) are in a key position to distribute knowledge and initiate concern for the environment. According to Russell Dalton (1994), we should acknowledge different types of green movements (e.g. traditional environmental protection movements and ecological movements), which all add up to a broad environmental concern. This chapter traces the different roots of activism in the Baltic Sea Region, emphasizing the different settings at hand (Jahn and Werz, 2002).

Cognitive-informational framework conditions refer to the conditions under which environmental knowledge is produced, distributed, interpreted and applied (Jänicke, 1997, p7). Knowledge is the first factor to be recognized. If no problem is perceived, there will be no public outcry for proper environmental politics either. In the second part of this chapter a comparison of public awareness in the countries surrounding the Baltic Sea is conducted based on World Value Surveys. Most Different System Design is used to test whether there is a north-western and south-eastern policy style, which would form the background conditions for environmental concern (Przeworski and Teune, 1970).

TRADITIONAL ENVIRONMENTAL ORGANIZATIONS: EARLY START IN THE NORDIC COUNTRIES AND GERMANY

Traditional environmental protection organizations emerged early in Nordic countries and quickly found their way to positions of influence. In Denmark, a Commission for Nature Preservation was established in 1905, later succeeded by the Nature Conservancy Association in 1911. The association was prominent on conservation questions (e.g. on the first nature conservation law of 1917 and later amendments), and it became involved in the political and administrative decision-making processes on conservation. The association's advocacy work also contributed to the plan for the Aquatic Environment in 1987. As one of the largest organizations in Denmark, it had around 250,000 members during the 1990s (Christiansen, 1996, pp59–60, 63–64).

Similarly, Norway also served as a power in the development of environmental movements in the Nordic countries. The Association for Nature Preservation (ANP) organized conservationists in 1914, while the Norwegian Mountain Touring Association, founded in 1868, held an early interest in nature as a source of recreation. During the 1930s, the social basis of the ANP was enhanced as preservation was felt to be too narrow a concept. With another new wave of mobilization during the late 1960s and 1970s, it was reorganized into the Norwegian Society for the Conservation of Nature, adopting a more ecologically oriented ideology and promoting more action-oriented initiatives (Jansen and Osland, 1996, pp180–182).

The Swedish Society for Nature Conservation, established in 1909, spent many years lobbying for a prominent role in environmental decision-making. From a rather narrow interest organization, it soon expanded and now counts 200,000 members. The society, along with the Swedish branches of the World Wide Fund for Nature (WWF) and Greenpeace, now claims 5 per cent of the population as paying members. Although a membership does not automatically require a deep commitment to the movement, many surveys show broad popular concern for the environment (Lundqvist, 1971, 1996, pp285–288).

The Finnish Nature Protection Association was founded in 1938. An ecologically minded wing was able to influence the organization, expanding the ideological scope of its work, and there was no need to establish another organization for the new forces at work. Today there is a network of regional and local branches of the Finnish Association for Nature Conservation, and a corresponding organization among the country's Swedish-speaking citizens, Nature and Environment, claiming nearly 40,000 members during the late 1980s. Furthermore, the consumers' movements, which are of considerable strength in all Nordic countries, have joined this traditional movement (Hermanson and Joas, 1996, pp129–130).

We conclude that these traditional organizations still enjoy remarkable support since the strength of popular movements in the Nordic countries was an important feature of public policy in the region. Early attempts to formulate environmental policies had a conservation approach. More ecological thoughts emerged along the way; but generally these were also adjusted to fit existing structures.

However, the quality of the environment is affected by general arrangements and conditions in society, such as the rate of economic growth, changes in industrial structures and technology. The change in production as well as the structure, size and environmental impact of different sectors (such as agriculture and the forestry industry) has shifted the focus of environmental concern. Accordingly, these broad environmental groups developed a popular base and a high and salient environmental awareness. Crucially, political systems were open to this influence.

NEW MOVEMENTS AND GREEN PARTIES IN A NEW POLITICAL ENVIRONMENT IN THE BALTIC SEA REGION

National political systems differ in terms of transparency and the ability to meet new challenges. The environmental movement has developed and channelled the political message in different ways around the world. A key aspect has been the mobilization of ecological environmental awareness, strongly influenced by issues concerning social justice and minority rights, while the political organization has been less prominent. However, any comparison of strategies must consider the context, the political climate and strategies of the political elite (integrative or repressive) towards challengers, prevailing divisions in society, and the balance of power between various political parties and factions (i.e. 'political opportunity structures'; see, for example, Kitschelt, 1986; Heijden, 1999). Green parties have persistently opposed categorization and labelling along the traditional left- or right-wing spectrum. Instead, they have created a new cleavage: ecology versus growth, in which, loosely interpreted, the former indicates a new post-materialistic stance and 'new politics'. Is this the case in the Baltic? We will see that a left–right political scale is hard to grasp in the Baltic Sea Region.

Green parties and their development have mainly been studied in Western Europe (Müller-Rommel, 1989; Doherty, 1992; Richardson and Rootes, 1995; O'Neill, 1997). There is now good reason to consider the former Eastern Bloc countries as the European Union (EU) is rapidly extending. Together, members of national Green parties form one Green Group in the European Parliament (see www.eurogreens.org for the results of the European elections during 10 to 13 June 2004). Over the years, this group has become a well-established actor which confronts the dilemma of attempting to represent alternative, decentralized policies in a thoroughly bureaucratic and hierarchical structure, such as the EU (Bomberg, 1998, p124). Today, this cooperation has been strengthened through the establishment of the European Federation of Green Parties (see www. europeangreens.org), which ran a joint Green electoral campaign for the European parliamentary elections in 2004. However, with Western Europe as a point of departure, it is important to remember that the political structures of Eastern Europe have been, and still are, very different – and this has also influenced the actions of the Greens (Jehlicka, 1994; Rootes, 1997). In this context, Touraine's (1987, p221) comments are particularly poignant: 'New social movements often have more influence as instruments of transformation of political life than as themselves'.

Policy developments in Germany were pushed by citizens' initiatives and the rise of the Greens. During the 1980s, Germany became one of the forerunners in environmental policy. The reasons for this may have been the strong mobilization of the environmental and anti-nuclear movements, as well as the establishment of Die Grünen in 1979. After 1994, Germany lost its progressive role due to severe economic constraints in the wake of German unification. A pragmatic alliance,

Bundnis 90/Die Grünen (the former being a loose coalition of citizen initiatives in former East Germany), provided new momentum to the movement (Jahn, 1994).

The history of the environmental movement in Estonia, as is the case in other Central and Eastern European countries, is closely related to the independence movement in the (early) 1980s. The activities by non-governmental organizations (NGOs) were constrained; but interest in nature was viewed as harmless. Environmental movements played an important role in these countries' emergence from the Soviet era, but then faced severe environmental challenges and times of economic recession (Eckerberg, 1994).

In Estonia, only a few NGOs were initially allowed to exist. The Estonian Green Movement was the largest. It gathered naturalists who recognized potential threats to the natural environment of Estonia during a transition period from a centralized system to a market economy and democratic society. As a result, the Estonian Fund for Nature (ELF) was founded in 1991. The Green party, Eesti Rohelised, was founded in 1989.

The Latvian environmental movement can be traced back to the foundation of the Latvian Society of Nature and Monument Protection in 1959. By the 1990s, it had become one of the two largest environmental organizations in Latvia, unifying 23 local branches. The new political process in the Baltic during the late 1980s and early 1990s spurred the development of a new wave of environmental NGOs, including the Environmental Protection Club.

Latvijas Zaļā Partija participated in the 1990 election and received one seat in parliament and two representatives in government. However, the election was not counted as a free election. In 1995 the party gained 6.1 per cent of the votes and 4 seats out of 100 in parliament.

In Lithuania, around the time of Perestroika, democratic reforms allowed for the creation of independent environmental and cultural heritage groups. When the new statutes governing NGOs and civil society groups were passed in 1987, the ecology movement was born and many NGOs began operating across the country. In the autumn of 1988, Lithuania's environmental NGOs established the Lithuanian Green Movement (LGM), an umbrella union of environmental clubs that played an important and successful role in environmental protection activities. During the first free elections in 1990, ten representatives of the LGM were elected to parliament. Later, the LGM moved in a political direction and created the Lithuanian Green party.

The period around 1980 can be seen as the first phase of the environmental movement in Poland. This was a period of great unrest across the country. Organizations committed to environmental concerns included the Ecology Club, set up in 1980, and the Green party, Polska Partia Zielonych, in 1988 (Waller and Millard, 1992). A second phase came around 1990. Before the election in 1991, Polska Partia Ekologiczna was established, as was Polska Partia Ekologiczna-Zieloni. Later on, the Green party appeared in coalition with the Alliance of the Democratic Left.

Although nation states have been the major actors in the field of environmental policy, the focus is shifting to transnational and regional levels, with the European Union as a principal player. The European Union, long striving for environmental consistency in policies among its old member states, pushed this issue in negotiations with its new member states; but key industrial and agricultural sectors were also at stake. Environmental organizations, in particular, felt that they had very limited input. EU funding, offered through the PHARE programme and the Regional Environmental Centre, aimed at bridging these difficulties. Intervening in the democratic processes in this way, such movements have become more professional and specialized. In Central and Eastern Europe, environmental movements are heading in the same direction as their Western counterparts (Hicks, 2004).

ENVIRONMENTAL ADMINISTRATION

The administrative response (i.e. the ability to adapt to new demands concerning the environment) became an issue in the Nordic countries during the early 1970s. Sweden got an early start as the Swedish Environmental Protection Agency was established in 1967. At the ministerial level, environmental issues were handled by the Ministry of Agriculture until 1986, when a new Ministry for Energy and Environment was established. In Denmark, the Ministry of Pollution was established in 1971, and the Norwegian Ministry of the Environment in 1972. The creation of the Ministry of the Environment in Finland was delayed due to a political and ideological dispute until 1983. This delay, among other reasons, led to the politicization of the environmental movement during the early 1980s. The environmental movement turned into Green political parties in Finland and Sweden. In Denmark and Norway, environmental concerns were, to a higher degree, channelled through existing parties rather than new Green parties. The nuclear question may have been decisive in mobilizing support for the new Green parties (Hermanson, 1999, 2001).

The German Federal Environmental Agency was established in 1974, while the Ministry of the Interior was in charge of environmental issues until the new Environmental Ministry was established in 1986.

Developments in Estonia, Latvia and Lithuania are presented in Chapter 5, and are not elaborated upon here. Information on these institutionalization processes is summarized in Table 4.1.

FROM INSTITUTIONALIZATION TO PUBLIC CONCERN

In summarizing these different institutionalization steps in Table 4.1, we notice that there is no clear divide between north-west and south-east. During the 1970s, environmental ministries were set up in Denmark, Norway, East Germany and Poland. National environmental agencies received an early start in Denmark, Sweden and West Germany. Here we have the frameworks for environmental concern and activities. What about public environmental concern: is it in line with administrative efforts?

A widespread interest in environmental matters is indicated by the upsurge of Green movements. Environmental protection organizations had, of course, an early start in the north-western area; but this is also the case in Poland. Green movements and Green party formation were a phenomenon throughout the Baltic Sea Region. In this context, the three Baltic countries have much in common. Nevertheless, the differences are as striking within a country set as between the north-western and south-eastern set. On Green representation in government, for example, we have countries such as Finland (1995), Germany (1998) and Latvia (1995).

PUBLIC ENVIRONMENTAL CONCERN IN THE BALTIC AND NORDIC COUNTRIES

Advanced industrial societies have experienced a shift towards more post-materialist value orientations (Inglehart, 1977, 1990). A widespread change in attitudes brought environmental issues to the political agenda during the early 1970s. Movements, gathering around different issues and influencing decisions on the local level, became more organized and challenged the national political decision-making processes. There was an early interest in environmental issues, especially in the Nordic countries. Initiatives to care about the environment took form in the attitudes of individuals, in the performance of movements and organizations, and even in the acts of governments.

Other European countries and their populations also developed an environmental awareness at an early stage. Throughout these societies, there was a concern about the state of the environment. Different issues were highlighted in different countries: protection of nature, nuclear energy issues, consumption patterns and lifestyles, to mention a few. Generally speaking, Germany ranked among the highest when considering concern for the environment. In 1982, for example, the government in the Federal Republic of Germany was caught by the sudden upsurge in public concern on issues such as the destruction of forests caused by acid rain. In the Swedish election of 1988, public opinion was formed, in part, by significant losses in the seal population. The rise of Green parties proved that the change of public attitudes influenced the political system, from the local level up to national parliaments (Müller-Rommel, 1989; O'Neill, 1997).

If this pattern is consistent for all developing societies, then it is clear that the stirrings of environmental consciousness and the subsequent evolution of environmental policy structures require fundamental and specific steps. A key question is whether there are sizable variations in the nature and scope of these steps in

Table 4.1 *Country-wise summary, with year for steps of institutionalization: First environmental/Green representation and establishment of environmental administration*

Steps of institutionalization: Environmental/ Green representation	Denmark	Finland	Norway	Sweden	FRG/GDR	Poland	Estonia	Latvia	Lithuania
Environmental protection organization	1905	1938	1914	1909	1899	1928	1988	1988	1988
Green movement	–	1980s	–	End of 1960s	1970s	Around 1980	1980s	1959	1980s
Green party	1983	1988	1988	1981	1979	1988	1989	1990	1990
First representation in parliament	–	1983	–	1988	1983	–	2007	1995	1990
First representation in government	–	1995	–	–	1998	–	–	1995	–
Ministry of the Environment	1971	1983	1972	1986	1986/ 1971	1972	1989	1993	1994
National Environmental Agency	1971	1986	–	1967	1974/ 1988	1980/91	–	–	2000

Source: National election results and reports, author's compilation

different countries. Does the pace of institutional development have an impact upon the nature of environmental governance? Do countries with a longstanding commitment to environmental protection have a more elaborate and developed environmental consensus?

The focus of the rest of this chapter is the institutionalization of environmental concerns in different settings (i.e. in Nordic countries and Baltic countries), representing different political cultures as well as different economic circumstances and ecological challenges. The results are examined country by country since this is the traditional level for actions aimed at environmental administration and policy-making. Here, Russia is included in the analysis.

THE METHODOLOGY OF THE STUDY

The similarity between Nordic countries means a comparative approach is frequently used in studying the region. Nordic countries started early in the development of administrative and legislative efforts, and they have generally kept a high profile in environmental affairs. Green attitudes have been documented in surveys, highlighting a popular base of environmental concern (Hermanson and Joas, 1999). Now, with the Baltic Sea Region in focus, the Governing the Common Sea (GOVCOM) project has raised the question of whether countries that have developed an environmental administration at their own pace and with a background of long evolution (compared to those that have undergone rapid transition) end up with common approaches to environmental governance. The question is if the same relationships between individual characteristics and high environmental concern, as well as action for a better environment, occur under different conditions. Is there a pattern for Denmark,[2] Finland, Norway, Sweden and Germany (West) that is different from that of Estonia, Latvia, Lithuania, Poland, and Russia? In order to enable this study, the comparative method should be discussed in more detail.

In comparative research there are basically two strategies: the Most Similar System Design (MSSD) and the Most Different System Design (MDSD) (Przeworski and Teune, 1970). While MSSD is used when comparing within a similar setting (e.g. the Nordic one), the latter method is used when units of research are as different as possible, but nevertheless show the same phenomenon that we wish to study. MSSD is suited for cases with variable relations at the system level, while MDSD can be used with variables at a sub-system level. What makes MDSD suitable in this context is the focus on variables below the system level (i.e. the attitudes held by individuals). The task is, then, to test one particular finding within a variety of systems – in this case, whether environmental concern in terms of environmentally sound lifestyles and Green activism are found in different settings in the Baltic Sea Region, and if the same individual factors lead to this environmental concern and readiness to act in different contexts.

Data used in this chapter is from World Values Surveys (WVSs), providing a systematically surveyed, comparable data set that is well suited to time period comparisons as well. The aim is to illustrate the current situation in order to understand how the differences appear in this set of countries: on the one hand, the north-western part consisting of the Nordic countries and Germany, and, on the other, the south-eastern countries consisting of the Baltic countries, Poland and Russia. Germany is still reported with two samples, east and west. The following section refers to the sample East Germany as a former country – in short, just a country among the others – although 'a sample of eastern Germany' would be more correct. Material from the third wave (1995–1997) of World Value Surveys will be used. The main question is whether we can find similarities among the respondents from the north-western countries compared with south-eastern countries, instead of an ad hoc distribution country wise according to different issues.

ASPECTS OF ENVIRONMENTAL CONCERN

A substantial number of studies deal with 'environmental concern': how people perceive environmental problems, to what degree problems are considered to be serious, and how they should be handled. Surveys of different social groups differentiated, for example, according to education, occupation, ethnicity and age examine respondents' concern for the environment. Shifts in concern over time, as well as differences between countries, have become a large area of interest, while survey questionnaires have become more standardized. However, the comparability of material is still a problem.

As mentioned above, the broad concept of 'environmental concern' includes a wide range of issues. It refers to awareness of environmental problems and support for environmental protection. It reflects attitudes, related understanding of, and behavioural intentions towards, the environment (Dunlap and Jones, 2002, p484). The authors point out that environmental attitudes, the term used foremost in empirical studies, are largely synonymous with environmental concern. Furthermore, they convincingly clarify the concept in examining a huge array of studies and tuning the definitions to key conceptual components. Particular issues or sets of issues, such as pollution, dominate. Another focus, drawn from theoretical constructs, is the way in which concern is expressed and translated into action. Generally defined:

> ... environmental concern refers to the degree to which people are aware of problems regarding the environment and support efforts to solve them and/or indicate a willingness to contribute personally to their solution. (Dunlap and Jones, 2002, p485)

The early postulated conflict between economic growth and environmental protection could be tested according to the statements in the 1995 World Values

Survey (WVS-95) shown in Table 4.2.[3] In the 1999 European Values Survey (EVS-99) there is no such statement, indicating that there might be other pressing priorities than environmental protection. Some countries are quite understandably more concerned with achieving a rapid improvement in their standard of living. Economic development is an obvious concern in order to avoid social unrest and to promote political stability. Environmental concern may, in the short run, be a luxury that the governments of these countries cannot afford, or are not willing to enforce if popular expectations are more in favour of growth.

Table 4.2 *The respondents' priority of environment or economy, country wise (percentage)*

Nation	Environment	Economy	Other	Don't know	Total
West Germany	42.7	39.8	12.5	5.0	100.0
Norway	62.5	32.6	4.2	0.8	100.0
Sweden	59.0	26.8	5.2	9.1	100.0
Finland	40.9	47.9	5.8	5.4	100.0
Poland	40.8	39.9	3.3	16.0	100.0
East Germany	32.7	51.0	12.4	3.9	100.0
Lithuania	27.0	47.1	2.4	23.6	100.0
Latvia	40.3	36.1	14.2	9.4	100.0
Estonia	41.3	43.6	7.5	7.5	100.0
Russia	44.6	31.2	3.7	20.5	100.0
Total	43.4	38.7	6.9	11.0	100.0

Source: World Values Survey (1995)

Strong support for environmental protection is found among the respondents in Norway, 62.5 per cent preferring to protect the environment, and in Sweden, 59 per cent. In Sweden, the support for economic growth is particularly low at 26.8 per cent. It is striking, though, that respondents in both East and West Germany, as well as in Latvia, come up with other solutions to a level of 12 to 14 per cent. Another interesting point in Table 4.2 is the amount of 'don't know' answers (an alternative never read out loud in the interviews) among Lithuanians and Russians, and, to a certain degree, also among Poles.

This analysis of the data, showing environment and economy at parity in terms of public concerns, would, of course, need further elaboration. The economic circumstances should be covered in more detail: both short fluctuations and the economic base of the systems, the current state of development and so forth. An unusually high unemployment rate of nearly 20 per cent and severe budget deficit problems in Finland at the time of the survey explain the Finnish tilt towards economic growth. Nevertheless, this outcome does not explain the division of attitudes between north-western–south-eastern attitudes.

A more fundamental comment would be that a single item hardly contributes evidence for environmental concern. Therefore, this chapter has created an index by adding two economy-related statements.[4] A reliability analysis of the scale (alpha item-total correlation) was run on both the whole material and country by country, showing that the variable 41 statement did not bring any additional explanation: it did not fit the scale. A dependent variable was created called 'willingness to pay'.

There are no clear-cut differences in this analysis (see Table 4.3) of mean values for respondents country by country; there seems to be an acceptance that a clean and healthy environment is costing money everywhere: interesting; but this is just a starting point. This makes up the first dependent variable 'willingness to pay' (WTP) to be analysed in terms of individual and cognitive factors in correlation with WTP. The hypothesis is that although different settings are at hand, the same variables will result in WTP.

Table 4.3 *Respondents' willingness to pay due to environmental reasons*

Nation	Mean value	Number of cases (n)	Standard deviation country wise
West Germany	5.5560	1017	1.23487
Norway	5.0288	1127	1.59312
Sweden	5.7022	1009	1.46935
Finland	4.8343	987	1.60238
Poland	4.7910	1153	1.65586
East Germany	5.1036	1009	1.25828
Lithuania	4.8082	1009	1.21993
Latvia	4.7813	1200	1.14533
Estonia	5.0127	1021	1.51345
Russia	5.1485	2040	1.37580
Total	5.0749	11,572	1.44362

Note: Scale: 0 (not willing to pay at all) to 8 (maximum).
Source: World Values Survey (1995)

Factor analysis on a series of questions concerning the respondents' life and activities during the last 12 months gave interesting enough support for two new variables, which this chapter identifies as:

1 environmental activism (V45 and V46);[5]
2 Green lifestyle (V42, V43 and V44).[6]

Table 4.4 shows that there is a difference in the extent of environmental activism running along the presupposed border between a north-western and a south-eastern area. Respondents from the latter area show remarkably low activity. It is clear that former East Germany is part of the north-western pattern in displaying a Green activity level of .34 on a scale of 0 (no activity) to 2 (have done both). Despite the division of two levels of activism, the hypothesis is that although different settings are at hand, the same variables will result in environmental activities.

Table 4.4 *Environmental activism among respondents*

Nation	Mean value	Number of cases (n)	Standard deviation country wise
West Germany	0.54	1017	0.709
Norway	0.45	1127	0.605
Sweden	0.42	1009	0.631
Finland	0.25	987	0.532
Poland	0.12	1153	0.378
East Germany	0.34	1009	0.568
Lithuania	0.05	1009	0.260
Latvia	0.08	1200	0.301
Estonia	0.07	1021	0.288
Russia	0.08	2040	0.308
Total	0.22	11,572	0.500

Note: Scale of 0 (no activity) to 2 (have done both).
Source: World Values Survey (1995)

The third dimension of environmental concern suffers from a lack of data on Estonia and Latvia, with V43 having no responses. In order to make the best of the data and to enable comparisons to be made, the new dependent variable was created as V42 and V43 for eight nations, as shown in Table 4.5. For the two nations mentioned, the variables V42 and V44 were used instead. This result is shown in Table 4.6.

Table 4.5 *Green lifestyles among respondents for eight nations*

Nation	Mean value	Number of cases (n)	Standard deviation country wise
West Germany	1.70	1017	0.554
Norway	1.43	1127	0.704
Sweden	1.71	1009	0.557
Finland	1.51	987	0.711
Poland	0.60	1153	0.723
East Germany	1.68	1009	0.599
Lithuania	0.51	1009	0.693
Russia	0.49	2040	0.706
Total	1.12	9351	0.860

Note: Scale of 0 (no activity) to 2 (have done both).
Source: World Values Survey (1995)

Table 4.6 *Green lifestyles among respondents for Estonia and Latvia*

Nation	Mean value	Number of cases (n)	Standard deviation country wise
Latvia	0.67	1200	0.724
Estonia	0.83	1021	0.748
Total	0.74	2221	0.739

Note: Scale of 0 (no activity) to 2 (have done both).
Source: World Values Survey (1995)

A pattern emerges distinguishing between a north-western and a south-eastern profile on Green lifestyles. It is clear that former East Germany is part of the north-western pattern, displaying a Green lifestyle scoring 1.68 on a scale of 0 (have not done) to 2 (have done both). Despite the division of two levels or profiles, the hypothesis is that although different settings are at hand, the same variables will result in a Green lifestyle.

INDEPENDENT VARIABLES AND HYPOTHESES

A number of studies, both theoretically and empirically oriented, as well as surveys within different contexts, have pointed out certain features leading to high environmental awareness. Drawing upon this rich literature, some hypotheses are presented and tested in the Baltic Sea Region context. According to Bennulf (1990, 1994), Dunlap and Jones (2002) and several other researchers, there are individual characteristics, based on basic demographic data, which make a difference (e.g. gender, age, education, occupation and class). The following sections start out with the first two since the other variables were difficult to use.

Age

Young people are more aware of environmental issues and will score higher on the variable of 'environmental concern' than older people. Three age groups will be examined according to this hypothesis: 18 to 39 (young); 40 to 59 (middle); and 60 upwards (old). The mean values for the whole sample are first mentioned, and then the results are commented upon country by country.

Willingness to pay

Respondents in the age group of 18 to 39 scored 5.20, while the middle-aged group received a mean value of 5.10, and respondents over 60 scored 4.85 – hence, the hypothesis is supported. Turning to the north-western group of countries, the hypothesis is not unquestioned in former West Germany, where both the

18 to 39 age group and the 60 plus age group received high scores: 5.56 and 5.58, respectively. Within the south-eastern country group, Estonian young and middle aged scored 5.07 and 5.12 compared to the 'old' respondents, with 4.64.

In eight countries the statement holds true: the 18 to 39 age group is more willing to pay for environmental improvements than the other groups. West Germany and Estonia, where differences across age ranges were slim, were the exceptions. There is no support for emphasizing the north-west–south-east differences.

Green lifestyle

In general, respondents aged 40 to 59 have more sustainable lifestyles than younger as well as older groups. This finding, based on the total sample, is highlighted within the north-western group. Sweden is the exception. In the south-eastern group, respondents from two countries break with the rest. Young people (compared to the other two age groups) in Poland and Russia express a way of life here recognized as a Green lifestyle.

In fact, a Green lifestyle at a dominant young age is found only in Poland and Russia. The general conclusion is that the middle aged in most countries score high on a Green profile of living. Geographic location may determine whether Green lifestyles are pursued or not; however, age seems to be a more likely variable. Area differences might explain only the level or frequency of Green lifestyles, with young to middle age being a more interesting variable to search for.

Green activism

The general figures on age give young respondents (aged 18 to 39) a slight lead in environmental activism compared to the middle aged. Then there is a gap down to a lower level of activism among old respondents. Germany, east and west, and Norway fit into this picture very well. For example, the mean values for West Germany are .60, .50 and .43 for each age group, respectively. However, when turning to the other north-western countries, this result diminishes. Swedish respondents score .40, .47 and .36, while their Finnish counterparts score .23, .24 and .31. In other words, there is no common feature according to age groups. The south-east country group is more coherent, with the middle aged showing most Green activism, though the level of activity is very low throughout all countries concerned.

Sex

Women are more concerned about the environment and will score higher on the variable of environmental concern than men. There are 5323 male respondents (46 per cent) and 6249 female respondents (54 per cent). Again, mean values for the whole sample are first mentioned, and then the results are commented upon country wise.

Willingness to pay

Women in all countries score higher on willingness to pay for the environment, with the exception of Poland. Here, male respondents receive a mean value of 4.82 and female respondents 4.77. Mean values may miss major differences within the groups. However, given that the survey populations are large and that standard deviation is in line with the other groups' results, no obvious explanation is at hand.

Green lifestyle

Women in all countries score higher than men on Green lifestyle, with the exception of Russia. While male Russian respondents score .51, female respondents score .47 (a minor difference between these values). Nevertheless, the differences expressed and displayed by gender in other countries are quite notable in demonstrating that women favour a Green lifestyle.

Green activism

On Green activism, again, women score higher than men; but on this aspect of environmental concern there are more exceptions. In West Germany, Lithuania and Estonia, the mean values are almost equal, with activism amongst men slightly more common. In Poland, the mean value is the same for both sexes: .12. With the hypothesis unchallenged in only six out of ten countries, it is the weakest point in the three dimensions of environmental concern. There is no support for the north-west–south-east division either, with respondents from West Germany making up an exception to the hypothesis together with Lithuanian and Estonian respondents.

Left–right orientation

Some cognitive characteristics have also proved to be related to high and salient environmental consciousness. The left–right orientation was tested: respondents with a left-wing orientation express more environmental concern than those with a right-wing orientation. Respondents were asked to place their views on a scale of 1 (= left) to 10 (= right).

The hypothesis held perfectly true in Germany (west and east), Norway and Sweden. Finns with a right-wing orientation were as active as left-wingers. The results, however, turn around for the south-eastern group. Nevertheless, the scale did not make much sense in those countries. With one quarter of the respondents (22 to 26 per cent) and half of the Russians (46 per cent) marking 'do not know', the self-placement should perhaps not be overly emphasized. It does, however, point out the difference in political scaling between the north-western group and the south-eastern group. A further comparison would perhaps not be correct here.

PERSONAL COMMITMENTS AND SUPPORT FOR
ENVIRONMENTAL MOVEMENTS

In the World Values Survey there is a classical question on whether or not the respondent is a member of voluntary organizations. The results on membership of environmental organizations are shown in Table 4.7.

Table 4.7 *Membership of environmental organizations (percentage country wise)*

Nation	Active	Inactive	Not member	Total
West Germany	3.0	8.8	88.2	100.0
Norway	0.9	5.8	93.3	100.0
Sweden	2.2	10.7	87.0	100.0
Finland	1.1	6.6	92.3	100.0
East Germany	1.2	3.7	95.1	100.0
Lithuania	0.4	1.8	97.8	100.0
Latvia	0.8	1.2	98.0	100.0
Estonia	0.4	2.0	97.6	100.0
Russia	0.4	1.2	98.3	100.0
Total	1.1	4.2	94.7	100.0

Source: WVS (1995)

Popular movements are a special characteristic of Nordic countries. Membership in voluntary organizations is high, although most of the members are inactive. The largest share of active members is found in the western part of Germany. The countries of the post-Socialist Bloc, perhaps with the exception of former East Germany, show a lower level of organizational activity. Poland is unfortunately missing. Table 4.7 strengthens the view that there are different opportunities in the political environments of the north-west and south-east.

CONCLUSIONS

In this chapter, the primary question was whether the same variables, despite different settings at hand, would initiate environmental concern. In translation of environmental concern into action, three variables were taken into account: willingness to pay out of environmental reasons; occurrence of a Green lifestyle; and engagement in environmental activities.

Generally, young age proved to be connected with willingness to pay. A Green lifestyle, on the other hand, is more pronounced among the middle aged. Still, this is a trend throughout the study. Area differences might explain only the level or frequency of Green lifestyles; support among the young and middle aged seems to follow the same pattern on this point. The third aspect, Green activism, is more concentrated in the south-eastern country group with the middle aged than in the north-western group. Still, the level of activity is very low throughout the south-eastern countries.

Gender is another variable that seems to be 'universal' in this context. Women, more often than men, express a readiness to pay higher prices and taxes if it would help to protect the environment. They are also more in favour of a lifestyle identified as Green in choosing household products that are better for the environment and recycling products rather than throwing them away. In terms of the third aspect of environmental concern, here measured as activism, female respondents are more active than men in six countries out of ten.

Environmental concern in the south-eastern part of the Baltic Sea Region is quite similar to concern elsewhere. Given the potential support (e.g. among young people and women) for environmental issues – in this chapter articulated as willingness to pay, Green lifestyle and environmental activism – there would be different ways of bringing the issue out and calling for political commitment to environmental policy-making. Different groups in these societies could mobilize to demand an effective policy. Political leadership could recognize the need for further integration of environmental protection and other consequences of economic activities when trying to promote the country's development. Or, in a broader context, a commitment to environmental protection could stem from actors outside of the country in question. This is perhaps the current situation. Different commitments are imposed in return for participation in joint systems – for example, the Helsinki Commission on the Protection of the Marine Environment of the Baltic Sea Area (HELCOM) and the European Union.

An example of a joint effort in this context is the Nordic Council, founded in 1952 as a forum for cooperation among the parliaments and governments of Denmark, Iceland, Norway and Sweden. In 1955, Finland joined the group. The Nordic Council initiated and set up advisory functions and encouraged Nordic cooperation. The Nordic Council of Ministers was established in 1971 as an organization for cooperation between the Nordic governments. Depending upon the agenda, the Council of Ministers meets in different guises. Accordingly, the Ministers for the Environment have prepared programmes of environmental co-operation between the Nordic countries (e.g. Nordic Programme for the Environment, 1989). Joint Nordic Action Plans (NAPs) cover pollution issues (e.g. NAP on Air Pollution, 1990, and NAP on Sea Pollution, 1990) and promote the development of new technology (e.g. NAP on Cleaner Technology, Waste and Recycling, 1990). This picture of Nordic cooperation was suddenly broadened in time for the Baltic system shift. Hence, new forms of cooperation and new modes of governance are the focus of the following chapters in this volume.

NOTES

1   Although not the focus of this chapter, it should be remembered that the local level is important and is outstanding in its competence in Nordic countries. Long traditions of local autonomy and democratic ruling give the municipalities a central role in public administration structures. Therefore, the entrance of Green politicians on the local arena strengthens environmental awareness since the local level and the municipalities are important for policy-making and implementation.
2   Not included in the 1995 World Values Survey.
3   (V41) Here are two statements that people sometimes make when discussing the environment and economic growth. Which of them comes closer to your own point of view:
    * Protecting the environment should be given priority, even if it causes slower economic growth and some loss of jobs.
    * Economic growth and creating jobs should be the top priority, even if the environment suffers to some extent.
    * Other answer (volunteered).
4   (V38) 'I would agree to an increase in taxes if the extra money were used to prevent environmental damage.'
    (V39) 'I would buy things at 20 per cent higher than usual prices if it would help to protect the environment.'
5   (V45) Have you attended a meeting or signed a letter or petition aimed at protecting the environment?
    (V46) Have you contributed to an environmental organization?
6   (V42) Have you chosen household products that you think are better for the environment?
    (V43) Have you decided for environmental reasons to reuse or recycle something rather than throw it away?
    (V44) Have you tried to reduce water consumption for environmental reasons?

REFERENCES

Andersen, M. S. and Liefferink, D. (eds) (1997) *European Environmental Policy: The Pioneers*, Manchester University Press, Manchester.

Andersson, M. (2002) 'Environmental policy in Poland', in H. Weidner and M. Jänicke (eds) *Capacity Building in National Environmental Policy: A Comparative Study of 17 Countries*, Springer, Berlin, pp347–373

Bennulf, M. (1990) 'Det gröna genombrottet', in M. Gilljam and S. Holmberg (eds) *Rött blått grönt: En bok om 1988-års riksdagsval*, Bonniers, Stockholm, pp142–163

Bennulf, M. (1994) *Miljöopinionen i Sverige*, Dialogos, Lund.

Bomberg, E. (1998) *Green Parties and Politics in the European Union*, Routledge, London

Christiansen, P. M. (1996) 'Denmark', in P. M. Christiansen (ed) *Governing the Environment: Politics, Policy, and Organization in the Nordic Countries*, Nord 1996, vol 5, pp29–102

Dalton, R. (1994) *The Green Rainbow. Environmental Groups in Western Europe*, Yale University Press, New Haven

Dalton, R. (2005) 'The greening of the globe? Cross-national levels of environmental group membership', *Environmental Politics*, vol 14, no 4, pp441–459

Doherty, B. (1992) 'The Fundi–Realo controversy: An analysis of four European Green parties', *Environmental Politics*, vol 1, no 1, pp95–120

Dunlap, R. E. and Jones, R. E. (2002) 'Environmental concern: Conceptual and measurement issues', in R. E. Dunlap and W. Michelson (eds) *Handbook of Environmental Sociology*, Greenwood Press, Westport, CT, pp482–524

Eckerberg, K. (1994) 'Environmental problems and policy options in the Baltic states: Learning from the West?', *Environmental Politics*, vol 3, no 3, pp445–478

European Values Survey (1995–1997, 1999) ICPSR version, edited by R. Inglehart et al, Institute for Social Research, Ann Arbor, MI, and Inter-University Consortium for Political and Social Research, Finnish Social Science Data Archive, Tampere, FL (2000)

Heijden, H.-A. van der (1999) 'Environmental movements, ecological modernisation and political opportunity structures', in C. Rootes (ed) *Environmental Movements: Local, National and Global*, Frank Cass, London, pp199–221

Hermanson, A.-S. (1999) 'Green movements and their political development in the Nordic countries', in M. Joas and A.-S. Hermanson (eds) *The Nordic Environments: Comparing Political, Administrative, and Policy Aspects*, Ashgate, Aldershot, pp91–110

Hermanson, A.-S. (2001) 'Politik och gröna partier i Norden', in E. Blomqvist (ed) *Från teknik till etik. Miljöförståelse i universitetsperspektiv*, Åbo Akademis förlag, Åbo, pp105–114

Hermanson, A.-S. and Joas, M. (1996) 'Finland', in P. M. Christiansen (ed) *Governing the Environment: Politics, Policy, and Organization in the Nordic Countries*, Nord 1996, vol 5, pp103–177

Hermanson, A.-S. and Joas, M. (1999) 'Comparisons and environments', in M. Joas and A.-S. Hermanson (eds) *The Nordic Environments: Comparing Political, Administrative, and Policy Aspects*, Ashgate, Aldershot, pp3–12

Hicks, B. (2004) 'Setting agendas and shaping activism: EU influence on Central and Eastern European environmental movements', *Environmental Politics*, vol 13, no 1, pp216–233

Huntington, S. P. (1965) 'Political development and political decay', *World Politics*, vol 17, no 3, pp386–430

Inglehart, R. (1977) *The Silent Revolution*, Princeton University Press, Princeton

Inglehart, R. (1990) *Culture Shift in Advanced Industrial Society*, Princeton University Press, Princeton

Jahn, D. (1994) 'Unifying the Greens in a united Germany', *Environmental Politics*, vol 3, no 2, pp312–318

Jahn, D. and Werz, N. (eds) (2002) *Politische Systeme und Beziehungen im Ostseeraum*, Olzog, München

Jänicke, M. (1992) 'Conditions for environmental policy success: An international comparison', *The Environmentalist*, vol 12, no 1, pp47–58

Jänicke, M. (ed) (1996) *Umweltpolitik der Industrieländer. Entwicklung – Bilanz – Erfolgsbedingungen*, Edition Sigma, Berlin

Jänicke, M. (1997) 'The political system's capacity for environmental policy', in M. Jänicke and H. Weidner (eds) *National Environmental Policies: A Comparative Study of Capacity-Building*, Springer, Berlin

Jansen, A.-I. and Osland, O. (1996) 'Norway', in P. M. Christiansen (ed) *Governing the Environment: Politics, Policy, and Organization in the Nordic Countries*, Nord 1996, vol 5, pp179–256

Jehlicka, P. (1994) 'Environmentalism in Europe: An East–West comparison', in C. Rootes and H. Davis (eds) *Social Change and Political Transformation: A New Europe?*, UCL Press, London

Kitschelt, H. (1986) 'Political opportunity structure and political protest: Anti-nuclear movements in four democracies', *British Journal of Political Science*, vol 16, pp57–85

Lundqvist, L. J. (1971) *Miljövårdsförvaltning och politisk struktur*, Prisma, Lund

Lundqvist, L. J. (1996) 'Sweden', in P. M. Christiansen (ed) *Governing the Environment: Politics, Policy, and Organization in the Nordic Countries*, Nord 1996, vol 5, pp257–336

Müller-Rommel, F. (1989) 'Green parties and alternative lists under cross-national perspective', in F. Muller-Rommel (ed) *New Politics in Western Europe: The Rise and Success of Green Parties and Alternative Lists*, Westview Press, Boulder, CO

O'Neill, M. (1997) *Green Parties and Political Change in Contemporary Europe: New Politics, Old Predicaments*, Ashgate, Aldershot

Przeworski, A. and Teune, H. (1970) *The Logic of Comparative Social Inquiry*, Wiley, New York

Richardson, D. and Rootes, C. (eds) (1995) *The Green Challenge: The Development of Green Parties in Europe*, Routledge, London

Rootes, C. (1997) 'Environmental movements and Green parties in Western and Eastern Europe', in M. Redclift and G. Woodgate (eds) *The International Handbook of Environmental Sociology*, Edward Elgar, Cheltenham, pp319–348

Touraine, A. (1987) 'Social movements: Participation and protest', *Scandinavian Political Studies*, vol 10, no 3, pp207–222

Waller, M. and Millard, F. (1992) 'Environmental politics in Eastern Europe', *Environmental Politics*, vol 1, no 2, pp159–185

World Values Survey (1995–1997, 1999) ICPSR version, edited by R. Inglehart et al, Institute for Social Research, Ann Arbor, MI, and Inter-University Consortium for Political and Social Research, Finnish Social Science Data Archive, Tampere, Finland (2000)

5

Environmental Governance in the Baltic States: Lithuania, Latvia and Estonia

Panu Kontio and Kati Kuitto

Introduction

The Baltic states regained their independence at the beginning of the 1990s when the Soviet Union was falling apart and several Soviet Republics became independent. The newly independent states, though subject to many institutional problems, shared a common vision for integration within Western structures such as the European Union (EU) and the North Atlantic Treaty Organization (NATO).

Despite similar goals, the policy and governance landscape with regard to the environment was different in each of these three countries. This historical development was possibly influenced by differences in cultural background, selection of partners for cooperation, ethnic structures and political climate.

Transition is a key feature of all three Baltic states. The time frame for changes was quite different compared to Western European and Nordic countries, where environmental policy and governance gradually developed in tandem with increasing environmental awareness over decades. The difference compared to, for example, Nordic countries was that awareness of environmental governance issues emerged suddenly as they embraced political and economic reform. This new awareness comprised environmental policy instruments, environmental management styles, institutional structures, research and monitoring methods, and involvement of civil society in environmental decision-making. Much of this knowledge was rapidly transferred to the countries in transition (among them the Baltic states) through massive aid and cooperation. However, one cannot say that the Baltic states came out of the darkness on environmental issues; in fact, the environmental movement was one of the key forces in the fight for independence, and even in Soviet times there were administrative structures for environmental protection.

The purpose of this chapter is to briefly describe the development of environmental administration and environmental policy in the Baltic states, and to understand some of the similarities and differences in this development. The causes of the differences are very complex – notwithstanding, we will suggest some explanations. We will make reference to the development of environmental policy and governance in Western European and Nordic countries. However, it would not be justified to try to use the Nordic consensus (corporative) model (e.g. Lundqvist, 1996) to describe and analyse development in the Baltic states since the starting point and the circumstances have been so different.

We have chosen Estonia, Latvia and Lithuania as cases for deeper examination of environmental policy and paid less attention to the other Baltic Sea countries since they have already been comprehensively studied. We have focused on two issues that are minimally regulated by EU legislation: administrative structure from the point of decentralization and land-use planning (spatial planning). These areas have been chosen as they are more likely to reveal country-specific differences than other policy fields that are tightly regulated by the European Union.

The following section describes the common starting points of the Baltic states for developing their environmental policy and administration. The third main section describes the development of administration and environmental policy in each of the Baltic states during the 1990s. The descriptive analysis is primarily based on environmental policy documents. Section four discusses decentralization and the role of land-use planning in connection with environmental impact assessments, and the Baltic states are compared to each other. Finally, section five brings in the Western understanding of environmental policy and discusses how well development in a country in transition fits into these schemes.

BACKGROUND TO ENVIRONMENTAL GOVERNANCE IN THE BALTIC STATES

The Baltic states were part of the Soviet Union from the World War II until the beginning of the 1990s. The Soviet regime dictated the economic system, the administrative structure and the legal framework. From this point of view, the Baltic states inherited a common pattern of environmental problems.

One of the prominent problems was inadequate wastewater treatment in cities, neglected waste management and production based on outdated technology. This resulted in badly polluted soils and stores of agricultural chemicals kept in open storages around the countries (e.g. Kontio and Karlin, 1999; Purhonen and Kontio, 2000), along with high levels of industrial pollution and heavily polluted air near industrial sites, as well as rivers and groundwater polluted by industrial and municipal wastewater discharges, inefficient use of natural resources (timber, minerals, fuels and water) and neglected nuclear safety (OECD, 1999, p23).

On the other hand, the centrally planned economy had some advantages as well: there were large areas of natural or nearly natural habitat due to restrictions of personal movement, consumer waste levels were lower than in Western Europe, and emissions by traffic were lower due to fewer vehicles per capita (OECD, 1999, p24).

Soviet-style governing and policy-making was very different from the Western style – for example, environmental information was kept secret, the public did not have opportunities to participate in environmental decision-making, local self-government did not exist, and key civil servants belonged to the privileged class. One legacy of this style of governance is that Central and Eastern European countries faced a range of institutional problems: poorly functioning judicial systems, a lack of local government financial and administrative autonomy, limited institutional capacity, and a lack of a civil service legislation (OECD, 1999, p63).

Right after restoration of independence at the beginning of the 1990s, the varied environmental problems mentioned above came to prominence. This co-incided with rapid industrial decline and a dramatic drop in both production and consumption. Together, these resulted in a decrease in the use of natural resources and a remarkable reduction in environmental pollution (Lithuanian Ministry of Environment, 2002). It was in this context that a new system of environmental governance was initiated.

The first direction was the renovation of the legal system as national govern-ments took more control over natural resources. Soviet laws needed to be replaced – however, abandoning a number of Soviet laws overnight left many fields of environmental policy unregulated. A great number of laws were therefore passed hastily, without allocating sufficient time for preparation. Under these circum-stances the legal Acts had quite a short life span, and turned out to be temporary. During the mid 1990s, the EU approximation maintained the momentum of revising legislation.

Second, the environmental administration needed to be rearranged. During the Soviet era, rules were made in Moscow even though Rinkevicius notes that 'the existing latent form of environmentalism managed to ameliorate and adjust them into local conditions' (Rinkevicius, 2000), and some state-level environmental administration already existed during Soviet times. On the eve of independence, Latvia and Lithuania established environmental committees subordinated to the parliament. Estonia founded the Ministry of the Environment in the immediate aftermath of independence. During the first years of independence, changes in the administrative system were, to a large extent, influenced by political instability. The terms the governments served were quite short and the political pendulum swung from right to left.

The third direction was to restore the municipal system. Democratically elected local governments were re-established; but for many years they lacked the capacity to manage the basic tasks that are normally assigned to municipalities in Western Europe. Lack of funds and economic capacity hindered their performance and kept them very dependent upon the central government. There were, however, some exceptions, such as the rich cities of Riga, Ventspils and Jurmala in Latvia, and Vilnius, Klaipeda and Kaunas in Lithuania. These cities were large and/or had some independent sources of fundraising (e.g. car tolls in Jurmala or oil transit revenue in Ventspils).

The fourth direction in which change in environmental governance took place was in reaching out to the international community since it was a matter of priority to stabilize the independent status of the Baltic states. By signing and ratifying international conventions on environmental protection, the states showed progress in this respect. Among the first international forums they attended were the Helsinki Commission on the Protection of the Marine Environment of the Baltic Sea Area (HELCOM), the Environment for Europe (Dobris and Lucerne) and the United Nations/Economic Commission for Europe processes. After the (velvet) revolutions in Eastern Europe, the Dobris ministerial meeting in 1991 was the first international event to address environmental problems in Central and Eastern European (CEE) countries. An assessment was prepared for the meeting, which showed data for the first time on the state of the environment in the CEE countries. Based on this assessment, it was decided that actions were needed. The Dobris meeting initiated the Environment for Europe process, which was manifested in Lucerne. The environmental ministers endorsed the Environmental Action Programme (EAP), which aimed at lasting environmental improvements, not only through investments alone, but by combining policy reforms and institutional strengthening with cost-effective savings (OECD, 1999, p26). Bilateral cooperation was also established during the early 1990s, and numerous aid projects transferred knowledge and helped with environmental investments. The environment was one of the first areas to receive assistance by Western countries and organizations for a number of reasons. The environmental movement was one of the main engines of democratic reform in the CEE countries. Transboundary and global effects were regarded as important, and taking into consideration that pollution reduction per unit was much cheaper in CEE than in Western countries, aid was an effective way of reducing transboundary pollution. It was also a unique opportunity to integrate environmental, economic and political development (OECD, 1999, p24). One of the factors facilitating aid projects was the big market for Western environmental protection technology as subsidized (tied aid) exports were possible without constraints. Perhaps the most significant steps were the Baltic states' applications for EU and NATO membership. EU membership, in particular, has had an enormous impact upon environmental governance throughout the region.

ENVIRONMENTAL ADMINISTRATION AND POLICY IN THE BALTIC STATES

Lithuania

The Republic of Lithuania declared its independence on 11 March 1990. The last Soviet soldiers left the country in 1993. A new constitution was approved in a referendum in October 1992. A parliamentary system with a president as head of state was introduced. A one chamber parliament, Seimas, with 141 members is elected for a four-year term (Lithuanian Ministry of Environment, 2002).

Environmental administration

In 1990, parliament established a sub-committee with a strong mandate to protect the environment. The committee had dual roles: as a policy-maker as well as an adviser to the plenary. The committee exercised its power on numerous occasions, most notably when it overruled the government's decision to build the Buttinge Oil Terminal not once, but three times (Raulinatis et al, 2001). The committee evolved into the Environmental Protection Department before being transformed into the Ministry of Environmental Protection in 1994.

In 1998, it was merged with the Ministry of Construction and Urban Development, which had been responsible for territorial planning (land-use planning). In the merger, the forestry sector from the Ministry of Agriculture was transferred to this new ministry, which was named the Ministry of the Environment. The sphere of authority of the ministry covers environmental protection, nature protection, housing and building, spatial planning and regional development.

The implementation of environmental legislation falls within the remit of eight regional departments within the ministry. Numerous subordinate bodies to the ministry were assigned to carry out research, development, monitoring and supervision tasks. One of the most important bodies was the Joint Research Centre, which was transformed into the Lithuanian Environmental Agency in 2000 (Lithuanian Ministry of Environment, 2000). The State Environmental Inspection Team was assigned to carry out the supervision of environmental legislation.

The Strategy for the Approximation in the Environmental Sector followed Lithuania's signature of the European Union Association agreement in 1998. It identified a lack of internal capacity to deal with EU legislation and specifically noted that extra staff of 50 individuals would be needed. This strategy also noted the importance of integrating environmental issues with other sectors, such as transport and energy, and recommended that staff with an environmental background (or education) should be hired in these sectors. The number of staff in environmental administration had been constantly rising; but in 2000 there was a setback as the annual budget for the Ministry of the Environment was cut by 40.4 per cent from 1999 to 2000 (Lithuanian Ministry of Environment, 2000). It was difficult to manage the growing workload since the state economy was, at the time, struggling with economic crises. This was mainly a consequence of Lithuania's attempts to try to achieve the EU recommendations of share of state budget compared to gross national product (GNP). For example, the ministry staff's salaries and working time was cut to four days a week, although the 'unofficial recommendation' was that they would still work five days a week, full hours.

Environmental policy

Environmental policy is often created incrementally, with small steps taken over time. On other occasions policy is initiated by longer, more strategic documents.

One of the early environmental policy documents in Lithuania was the Resolution on the Organization of Integrated Monitoring and Establishment of a Factografic Information System EKOLOGIJA for the Status of the Environment in 1989. The Nature Protection Committee of (Soviet Republic) Lithuania decided to take action to improve understanding of the environmental situation in the country. During the Soviet era, most of the information concerning the environment was classified. Mikhail Gorbachev's *glasnost* (openness) policy during the late 1980s rendered it possible to make environmental information public, and his administrative reform *perestroika* policy made it possible for the Soviet Republics to establish their own institutions and organizations. Although many scientific organizations had been collecting environmental information before these new policies, this information was not available in an integrated form, neither had it been accessible to the wider audience. The work accomplished in the frame of this new monitoring system led to a publication on the state of the environment, which we will describe later.

In 1992, the Law on Environmental Protection (LEP) was approved. It lays down the basic principles of environmental protection. Its main objective is to achieve an ecologically sound and healthy environment upon which human activities have as little negative impact as possible, and to maintain Lithuania's typical landscape and its biodiversity (UN, 1999). The LEP is a framework law that forms the legal basis for enactment of other environmental legislation. A number of legal Acts were prepared and adopted, although we will not deal with them in detail in this chapter.

Parallel to the Law on Environmental Protection, the first National Environmental Protection Programme (NEPP) was prepared and approved in 1992. It addressed the major environmental problems in priority order and defined measures to solve them.

The first State of the Environment Report was prepared in 1993 in the Lithuanian language. For the 1995 English version, data from 1994 were included and some modifications were made (Environmental Protection Ministry of Lithuania, 1995). The report was prepared and published by a commission founded by the Environmental Protection Department, later the Environmental

Protection Ministry, to coordinate the implementation of the new 1993 Ecological Monitoring Programme. This commission consisted of the representatives of the department, as well as of scientific and research institutions. The objective of the report was to present the monitoring information covering all fields of the environment, except for the human health (environmental aspects) monitoring data, since the responsibility for monitoring human health was the task of healthcare institutions.

The report describes the objectives of the monitoring system, the system itself, the responsibilities of different institutions in data collection and, finally, information on the state of the environment. The environmental components that are included are air, water, soil, vegetation, wildlife, the landscape, wastes and radionuclide contamination. However, the level of accuracy in describing these components varies: in some areas, long-term monitoring data were available, while in others, time series covered only a few years.

The report concludes that during the early 1990s, environmental pollution had markedly decreased; this trend was due more to a general economic decline than to implementation of environmental protection measures. The main worry during this time was that when the economic recession ceased, environmental protection (i.e. pollution control) and other policy instruments would be in place to manage the new circumstances of free entrepreneurship and market economy.

The vision of EU membership had much influence on the Lithuanian Environmental Strategy (LES) and Action Programme, which was prepared in 1995 and approved in 1996. Most of the measures of the previous 1992 strategy had been implemented, and objectives had changed with EU integration. The English version of the publication contains the main statements of the Lithuanian version, which consists of three volumes: *Strategy Motivation*, volume 1; *Strategy Methodology*, volume 2; and *Action Programme*, volume 3. The strategy was adopted as an addendum to a parliamentary resolution.

The document briefly describes the state of the environment, sets environmental protection goals by sectors, discusses priorities, states the principles that are considered in the strategy, describes the instruments necessary for implementing the strategy, and finally outlines an action programme that describes the measures to be taken. In the conclusion of the LES, further actions for implementing the strategy are defined. These include annual action plans, financial calculations, selection of programme implementation partners, and measures to establish control mechanisms for implementing LES action programmes and use of funds.

The revision of the LES was to be done annually to consider economic development, changes in laws, environmental information, EU requirements, obligations arising from international agreements, funds allocated for environmental protection, and the needs of society. However, no English language updates have been produced.

The final step to get rid of Soviet remnants in the legislation was taken during the late 1990s. In 1997 the Law on Invalidating the Legislation Adopted before 11

March 1990 on the Territory of the Republic of Lithuania ended the era of Soviet legislation in Lithuania. It stipulated that all former Soviet regulations would be void as from 31 December 1998. The influence of the Act on environmental protection was that environmental standards ceased to exist. At the same time, the adoption of new (EU) standards was still under way, although many new mechanisms for regulating the environmental issues had already been introduced.

The EU was not the only international community that had a stake in the environmental policy of Lithuania. Joining international conventions, especially under the United Nations Economic Commission for Europe, shaped both policy and legislation in fundamental ways.

In parallel with signing and ratifying conventions, the Lithuanian environmental administration participated in the environmental performance review (EPR) of the United Nations Economic Commission for Europe (ECE) Committee on Environmental Policy during the late 1990s. Lithuania's EPR was included in the committee's work programme in January 1997. The report was submitted for evaluation by the Committee on Environmental Policy in May 1998 and was published in 1999.

Although the review was performed by external reviewers under supervision of the ECE, it can be considered as a domestic policy initiative in Lithuania since the concept and implementation are, to a great extent, based on the willingness and cooperation of the Lithuanian Ministry of the Environment. In the first part, the report reviews the framework of environmental policy and management: legal instruments, integration of economic and environmental decisions, introduction of cleaner technology, spatial planning, and international cooperation. The second part focuses on managing pollution and natural resources: air, water, nature and waste management. The third part discusses economic and sector integration: human health and the environment, environmental concerns and energy, and environmental concerns and agriculture and food processing.

During the mid 1990s, it became more apparent that Lithuania aimed for membership of the EU. A White Paper was signed with the EU in 1994, which defined the goals of harmonization of the regulatory system and integration within the single market. It set the agenda for transposition of EU requirements into Lithuanian legislation. This was followed by the Association Agreement in 1996, which outlined the needs and actions for transposition. The application for membership was submitted to the EU in 1995.

In 2000, the negotiations on Lithuania's membership began. In the negotiations it became apparent that the transposition of EU environmental legislation would be mostly done by the date of accession, with very little derogation to the transposition dates. Based on these conditions, Lithuania became an EU member state on 1 May 2004.

Latvia

The Republic of Latvia restored its independence on 5 May 1990 as the Parliament of Latvia (Soviet Republic) annulled its annexation to the Soviet Union and re-established the constitution of 1922. A referendum on independence was passed in March 1991. Latvia is a parliamentary democracy with a unicameral parliament: the 100-seat Supreme Council (*Saeima*). The president, who is the head of state, is elected by this body for a three-year term and is advised by a cabinet of ministers. The prime minister, who is the head of the government, is appointed by the president (Encyclopedia.com, 2002).

Environmental administration

In 1988 the State Environmental Protection Committee of the Soviet Republic of Latvia was founded. The '*perestroika*' policy gave the Soviet Republics the possibility of establishing their own institutions. After restoration of independence, the Environmental Protection Committee soon followed, which was established by the Law on the Republic of Latvia Environmental Protection Committee, approved in 1990. It was assigned responsibility for the execution of environmental protection and use of natural resources. The committee was directly subordinated to parliament. Eight regional environmental boards, subordinated to the committee, were responsible for implementing environmental protection on regional and local levels (*National Report of Latvia to UNCED*, 1992).

In 1993, the Ministry of Environmental Protection and Regional Development (MEPRD) was founded, based on the former Environmental Protection Committee. It was also given responsibility over regional development and land-use planning, which were new functions not taken care of by any other institution. Functions of construction, housing and municipal services were transferred from the previous Ministry of Architecture and Construction (NEPP, 1995). In 2003, regional development and land-use planning issues were delegated to the newly established Ministry of Regional Development and Local Governments, and the ministry was renamed the Ministry of the Environment.

Several subordinated bodies to the ministry were constituted over time to carry out implementation of environmental requirements. These numerous institutes have since been merged. At the beginning of 2005, all eight regional environmental boards, the Maritime Research Institute and the State Environmental Inspection were merged under the name Latvian Environmental Service, responsible for issuing permits and supervision. The Latvian Environmental Agency, the Hydrometeorological Institute and the Geological Survey were merged into the Latvian Environmental, Geological and Meteorological Agency. The State Environment Bureau is responsible for environmental impact assessments (EIAs) and is an appeal body for environmental permits.

Environmental policy

Reform of legislation and formulation of environmental policy started earnestly right after restoration of independence. In this section, we review some of the main legal acts that have relevance for environmental protection and some of the most salient environmental policy documents.

The 1991 Constitutional Law of the Republic of Latvia on Human and Civil Rights and Duties describes the basic rights of people. From an environmental point of view, it lays down the principles of right to appeal to the court, right to information, right to express opinions, and freedom of assembly and protest (Michanek and Blumberga, 1998).

The law on environmental protection was approved in 1991. It contains the principles and main tasks of environmental protection, is wide in scope and has a generally broad approach to environmental issues. It contains the basis for applying different legal instruments, rights of individuals, different environmental protection tools, institutional arrangements, competencies, enforcement mechanisms and sanctions (Michanek and Blumberga, 1998).

One of the first environmental policy documents intended for international forums was the 1992 national report of Latvia to the United Nations Conference on Environment and Development (UNCED) in Rio de Janeiro. It provides background information about the country, briefly describes the environmental situation in Latvia, and goes through the economic activities sector by sector, detailing their environmental problems. It defines the policies and legislation, the institutional framework, and sets goals and targets for environmental protection. It also briefly describes the Latvian State Environmental Programme, Environmental Protection, and international cooperation.

The national report was followed by the National Environmental Policy Plan (NEPP), prepared and approved in 1995. A great number of participants representing other ministries, regional authorities, non-governmental organizations (NGOs) and academia attended the workshops and contributed to the preparation of the policy plan. The policy plan sets out environmental policy goals for the coming decades, as well as the principles adhered to while developing strategies. It also lists the resources and mechanisms that may be used for implementing policies.

The initiation of the plan was inspired by the conclusions of UNCED in Rio de Janeiro in 1992, the Dobris conference in 1989, and the Environmental Action Programme for Central and Eastern Europe in Lucerne in 1993. The policy plan gives a short overview of the environmental situation in Latvia, sets goals and principles, and describes the prerequisites of environmental policy. The available policy instruments are described based on the situation at the time and plans are introduced for the near future to develop them. Priorities are set to tackle environmental problems and, finally, some proposals for the development of an action programme are provided.

The National Environmental Action Programme, as outlined in NEPP, was approved in 1996. The programme summarizes actions corresponding to the priority environmental problems addressed by NEPP.

The policy plan and action plan were followed by a great number of more specific plans and programmes during the latter half of the 1990s. One of them was the National Baltic Sea Protection Programme of 1996, the main programme in the field of environmental protection. It constitutes several sub-programmes for different sectors of environmental protection, such as the 1996 National Programme of Hazardous Waste Management, the 1997 state programme Water Supply and Wastewater Treatment in Small- and Medium-Sized Towns of Latvia (800+) and the 1997 National Municipal Solid Waste Management Strategy – Programme 500.

These sub-programmes attempted to improve the legislation, build added capacity and develop technology. In addition, they focused on increasing administrative resources. For example, the National Programme of Hazardous Waste Management included the establishment of the Programme Implementation Unit; the founding of necessary institutions for managing hazardous wastes; the training of authorities; the study of incineration possibilities; the establishment of hazardous waste storage and landfill; and renewing the legal basis for hazardous waste management.

Another important programme was the Water Supply and Wastewater Treatment in Small- and Medium-Sized Towns of Latvia (800+), an investment programme that was developed during 1995 and 1996 by a project with the same name. The aim of the programme was to make an inventory of the existing situations; to develop a strategy for renovation and further development of water services management in small- and medium-sized towns; and prioritization of investments and development of financing patterns. In practical terms, the programme helped (and still helps) the municipalities to prepare investments, assists in arranging financing, and organizes management of other activities.

A third sub-programme was the National Municipal Solid Waste Management Strategy – Programme 500, which sets the direction for developing municipal waste management. The main task of the programme is to reduce the negative impacts of illegal dumping and dumpsites and to increase recycling. At the time of the strategy there were more than 500 dumpsites in Latvia. The target was to establish 10 to 12 regional sanitary landfills to serve the whole country. For this purpose, the aim of the programme is to facilitate the establishment of inter-municipal waste management organizations in order to build and manage these landfills.

These are only some examples of policy documents that were prepared at the end of the 1990s, and were followed from 2000 on by several strategic documents, including the Biodiversity Strategy, the Climate Change Strategy and the Sustainable Development Strategy.

While adopting a number of national programmes, Latvia has also been active in international forums. Several international conventions in different fields of environmental protection were signed and ratified. The most important organizations in this respect are HELCOM and the United Nations Economic Commission for Europe. The national environmental policy was naturally influenced to a great extent by these forums and conventions.

During the mid 1990s, it was apparent that Latvia would aim to join the EU. A White Paper was signed with the EU in 1995, defining the process for bringing Latvia in line with EU law and setting the agenda for transposition of EU requirements into Latvian legislation. This was followed by an association agreement, signed in 1995, which came into force in 1998. It outlined the needs and actions for transposition, and in 1995 Latvia submitted its application for EU membership.

EU membership negotiations began in 2000. Similarly, as with Lithuania, the transposition of EU environmental legislation in Latvia was largely achieved by the date of accession, with little derogation. Latvia became an EU member state on 1 May 2004.

Estonia

The Republic of Estonia regained its independence on 20 August 1991 as the Supreme Soviet of the Republic of Estonia passed the decision on Estonian independence. A new constitution, based on the antecedent constitution of 1918, was adopted by a referendum in June 1992. The Republic of Estonia is a parliamentary democracy with a unicameral legislature, the *Riigikogu*, and a president as head of state. The 101 representatives of parliament are elected under proportional representation every four years. The country is divided into 15 administrative counties (Lagerspetz and Maier, 2002; Estonian Government, 2005).

Environmental administration

An embryonic environmental administration was set up in Estonia about two decades earlier than in the other Soviet Republics, making Estonia a forerunner in environmental matters within the Soviet Union. A governmental body for nature protection existed as early as 1957. In 1962, this body was renamed Forestry and Nature Protection Governance and Ministry for Forestry and Nature Protection in 1966 (Finnish Ministry of the Environment, 2004, pp6–7). In the course of state transformation, a Ministry of Environment was set up in 1989. The responsibilities of the ministry include the protection of air; the management and protection of water, mineral resources and nature, including forest, fish and game animals; and the management of waste (UN, 2001, p5). Previously, spatial planning was also part of the ministry's tasks; but since 2004 land-use issues have been handled by the Ministry of Interior (Peterson, 2004; Estonian Ministry of the Environment, 2005). The Ministry of Environment is subdivided between four deputy secretary generals dealing with environmental administration, environmental management, international cooperation and nature conservation and forestry. There are an additional 15 environmental departments at the county level working under the mandate of the Ministry of Environment.

During the 1990s, the environmental administration underwent several structural changes. Prior to 1991, the Ministry of Environment had no offices at the local level. Environmental departments within county governments were responsible for environmental management and protection measures at the regional level. After 1993, these environmental departments were administered by the Ministry of Environment, but still financed by local government budgets. This financing structure made it difficult for the Ministry of Environment to ensure the implementation of national environmental strategies according to the priorities set up at the national level. In 2000, the environmental administration was reorganized and county-level environmental authorities were put under the direct responsibility and budgeting of the Ministry of Environment. The 15 regional authorities at the county level (the county environmental departments) now deal with EIAs; issue permits for natural resource use, pollution discharge and waste management plans; and also deal with the management of local and regional protected areas (UN, 2001, p19; Peterson, 2004).

New units with more staff specialized in environmental law were established after 1999 in the Ministry of Environment as there was an increasing need for more capacity for dealing with legislation and strategic planning in the course of the EU accession process. Unlike many other accession countries, no special integration unit was established within the Ministry of Environment. Various departments were themselves concerned with the adjustment of EU directives in their fields of activity, the adjustment process then being coordinated by the international cooperation department (UN, 2001, p19).

The enforcement of environmental law is carried out by the Environmental Inspectorate, which is headquartered in Tallinn in association with seven regional departments. Prior to the restructuring in 2000, two separate inspectorates – the Nature Protection Inspectorate and the Marine Inspectorate – were responsible for law enforcement, and the inspections were carried out by the staff of the county environmental departments. After the restructuring, the authorizing and enforcement mechanisms of the environmental administration became distinct from each other.

Besides the ministry and its county departments, the environmental administration in Estonia consists of research centres and institutes concerned with meteorology, geology, forest and land management, and environmental information. These bodies are under direct supervision of the Ministry of Environment (UN, 2001, p20).

Environmental policy

Environmental problems were already widely acknowledged in Estonia prior to the transition at the beginning of the 1990s. The alarming state of the waters, air

pollution (particularly in north-eastern Estonia), as well as the need for investments and international cooperation were documented in analyses during the late
1980s (Nordic Project Fund, 1989). Environmental movements formed part of
the democratic reform and independence forces prior to 1990, and environmental
awareness amongst the people carried on into the first years of the 1990s.

Environmental protection was incorporated in the new constitution of 1992.
Soviet laws were nullified in the years following independence and there was
a pressing need to develop a new body of environmental legislation to ensure
compliance with international agreements, while also entering domestic environmental protection. The Nature Protection Act adopted in 1990 provides an early,
but still valid, framework of principles and objectives of Estonian environmental
policy. Together with the 1995 Act on Sustainable Development, it constitutes
part of the environmental code drafted in 1998. By 1995, most of the relevant
environmental laws were passed; but many of them were still too broad and
lacking clear instructions for implementation. During the last ten years Estonia
has, nevertheless, adjusted its legislation, most notably in order to bring it in line
with the EU *acquis communautaire*. The current challenge for Estonia is to ensure
the implementation and enforcement of the laws (UN, 2001, p22).

The current environmental policy in Estonia is based on the National Environmental Strategy (NES), adopted by parliament in 1997. It contains ten principal
policy goals ranging from promotion of environmental awareness, introduction
of clean technologies and reduction of pollution to better use of natural resources
and improvement of the quality of the built environment (National Environmental
Strategy, Estonia, 1997). The policy goals formulated in the NES were compiled
into a National Environmental Action Plan (NEAP) in 1998. The first NEAP was
compiled in close cooperation with the EU and introduced a three-year action
plan with more than 650 projects financed by diverse governmental, private and
international sources. In 2001 the second NEAP was issued. The second action plan
was entirely managed by Estonian experts; but the broad involvement of potential
stakeholders was even more evident than during the formulation of the first NEAP.
The draft of NEAP 2 was also exposed to the public via the internet throughout
the formulation process, thus representing the new policy of open, participatory
legislation and public administration established in Estonia (UN, 2001, p11).
As some of the goals of the NES were achieved, some became obsolete and new
goals (such as protection against radiation and strategic planning issues) emerged.
During the elaboration of NEAP 2, a new National Environmental Strategy was
initiated and is currently being compiled (UN, 2001, p12; Peterson, 2004).

In order to incorporate the EU directives into domestic legislation and to
achieve the environmental goals set by the accession requirements, the National
Programme for the Adoption of the Acquis (NPAA) was implemented in 1998.
However, the objectives and implementation fields of the NPAA were narrower
than those of the NEAP, concentrating more on the adjustment of the legislation to
the *acquis communautaire* of the EU than on concrete environmental measures.

As the Act on Sustainable Development was adopted in 1995 (amended in 1997), Estonia became the first of the CEE countries to introduce a law addressing sustainable development issues. A National Commission on sustainable Development (NCSD) was set up 1996. Until now, however, the commission has played only a modest role in the legislation process. A National Sustainable Development Strategy, Estonian Agenda 21, was initiated by the government in 2001, and the Estonian National Strategy on Sustainable Development (Sustainable Estonia 21) was approved by parliament in September 2005 (Estonian Ministry of Environment, 2005). Participation in the Baltic Agenda 21, instigated in 1996, is also part of Estonian involvement in sustainable development in the Baltic Sea Region (UN, 2001, p13; Vaht, 2001; Estonian Ministry of Environment, 2002).

The process of privatization and the structural changes that occurred during economic transformation have also had consequences for environmentally friendly and socially compatible land-use and spatial planning. Land privatization is managed under the 1991 Land Reform Act. The Planning and Building Act, adopted in 1995, introduced spatial planning as a common practice and determined the Ministry of Environment to be responsible for land-use issues. The Act designated each county to devise its own spatial plan within three years and requested the counties to impose an action plan for implementing the land-use plans. By 1998, most of the counties had worked out a spatial plan in cooperation with the municipalities. The Planning and Building Act also asked the counties to introduce a strategy for the social and economic development of their territory. An overall strategy for spatial planning, the National Spatial Plan ESTONIA 2010 (EESTI 2010), was then compiled in 2000 by the Ministry of Environment. It integrates environmental, economic and social aspects of spatial planning. Since 2000, however, the role of the Ministry of Environment in the field of spatial planning has been diminishing – building issues were transferred to the Ministry of Economy in 2000, and since 2004, land-use issues are among the responsibilities of the Ministry of Interior (National Spatial Plan ESTONIA 2010; UN, 2001, pp13–14).

A great part of Estonian environmental policy takes place within the framework of international and regional cooperation. External financial and technical assistance and expertise have played an important role both for the development of environmental administration and environmental policy in Estonia during the 1990s. The cooperation with Finland has been particularly remarkable. Unofficial cooperation between Finnish and Estonian environmental NGOs already existed during the 1970s. In 1991, an intergovernmental agreement on environmental cooperation between Finland and Estonia was signed. This was the first international treaty Estonia signed after regaining its independence (Finnish Ministry of the Environment, 2004). Other states, most notably Sweden and Denmark, as well as many international and regional stakeholders such as the EU, the International Monetary Fund (IMF) and the European Bank for Reconstruction and Development (EBRD), have also been involved in environmental projects in Estonia, especially during the first years of the transition (European Bank for Reconstruction and Development, 1995; European Parliament, 1998).

One of the most important international environmental obligations of Estonia concerns the Baltic Sea, essentially under the HELCOM Convention. To date, Estonia has also ratified most of the major international and regional environmental conventions (Estonian Institute for Sustainable Development, 2001; UN, 2001, pp17–18).

SAME GOALS, DIFFERENT PATHS

Despite similar starting points and the common goals of joining the EU and NATO, the development of environmental administration and the adoption of environmental policy tools differ in Estonia, Latvia and Lithuania. The main bulk of environmental legislation was developed in all of these countries as a result of pressure from the EU. However, one of the areas less regulated by the EU is land-use planning, an area in which it was possible for the applicant countries (as well as existing member states) to choose their own way. Another less regulated area is the structure of environmental administration. In this section, we discuss these two areas: first, environmental administration from the viewpoint of decentralization of decision-making powers; and, second, the role of land-use planning as an environmental policy tool.

Decentralization of environmental governance

Every administration has its own pattern, organizational structure and culture that determine how policy is made. This fragmentation can be divided into vertical and horizontal components. Vertical fragmentation concerns how the responsibilities are distributed among different governmental levels: central, regional and local. By contrast, horizontal fragmentation deals with the distribution of these responsibilities among different authorities at each level (Lundqvist, 1996).

This sub-section describes how environmental authority in Lithuania, Latvia and Estonia is distributed within the vertical and horizontal dimensions of administration and how this has changed over time. Two trends in the development of state environmental administration can be found in all Baltic countries, the first being the separation of environmental administration from county organization, and the second the decentralization of decision-making powers from ministry to regional bodies. At the same time, municipalities are struggling with insufficient resources in arranging municipal environmental services – however, new arrangements are emerging in the field of inter-municipal cooperation.

The vertical fragmentation can be separated into two main dimensions. The first is the structure of the state environmental administration and the second is the division of tasks between the central government and local authorities. The first can be evaluated according to how centrally the state administration manages its tasks: are the decision-making powers concentrated in a central body (ministry or central agency subordinate to ministry) or distributed to the regional authorities that have decision-making powers of their own? The second can be evaluated based on the nature of the legally binding decisions of the local authorities in environmental matters and by budgetary independence of the regional and local authorities.

The horizontal dimension also has two sides to it: first, how the tasks are divided within the environmental sector between different agencies and research institutes; and, second, to what extent the management of environmental issues have been integrated within the other sectors, such as transport or energy. An indication of the first one could be the number of specialized institutes within the environmental sector and their mandate for making decisions, and the second could be defined by the number of environmental units or civil servants in the other sector organizations.

The 1990s brought along a series of changes both in the horizontal and vertical division of tasks, and resulted in reforms in administrative structures in all Baltic states. New institutions emerged and the number of bodies subordinate to the ministries of environment grew. In addition, the regional state environmental administration experienced changes. Local authorities were re-established right after independence and were restructured over time. Moreover, the role of local authorities evolved as they were assigned tasks in environmental decision-making.

State environmental administration

The ministries of the environment naturally lead policy-making and the implementation of government policies. Under the ministries, the subordinate agencies, research institutes and regional bodies translate the policy into concrete action. Despite the theoretical difference between horizontal and vertical fragmentation, these cannot, in practice, be dealt with separately. From the vertical perspective, distribution of decision-making powers is determined by the role of central agencies under the ministry and their relationship with regional authorities. For some issues, the system has three tiers of responsibilities: ministry, agency and regional bodies. For other issues, the regional body acts directly under the ministry.

In Lithuania, the regional environmental authorities are departments within the ministry structure. As a consequence, the ministry has a tight grip on decisions since decision-making is an internal matter of the ministry. Regional departments have their offices in local municipalities to perform the supervision tasks. The trend has been to give more authority to the regional departments through their approval of EIAs. When the law on EIAs came into force at the beginning of 2000, the regional authorities prepared the judgements; but the final decision was taken by the central body of the ministry. Within a few years, decision-making was transferred to the regional departments, and today the ministry makes the decisions only in exceptional cases – for example, concerning nuclear power, transboundary EIAs or trans-regional EIAs (Raulinaitis, 2005). On the other hand, inspection, led by the State Inspectorate, which forms an intermediate body between the ministry and the regions, works independently. This may reflect the principles of inspection – namely, externality and impartiality. Lithuania managed little by little to bring together most of the subordinate institutes into one body, called the Joint Research Centre, which subsequently became the Lithuanian Environmental Agency in 2000.

In Latvia, the implementation of environmental legislation and government environmental policies became a task for the regional environmental boards, which are responsible for issuing environmental permits for smaller projects. However, larger projects of national importance are still the preserve of central institutes such as the Latvian EIA Bureau or the Permit Board. The situation is changing continuously – today the EIA bureau is the Latvian State Environment Bureau responsible for EIAs and the appeal body for integrated environmental permits. At the beginning of 2005, all eight regional environmental boards, the Maritime Research Institute and the State Environmental Inspection were merged into the Latvian Environmental Service responsible for issuing permits and supervision. At the beginning of 2005, the Latvian Environmental Agency, the Hydro-meteorological Institute and the Geological Survey were also merged into the Latvian Environmental, Geological and Meteorological Agency. All of these changes shift the balance of implementing environmental policy to subordinate bodies from the ministry, which does not have a strong formal position in instigating environmental policy, and thus concentrates on making environmental policy and preparing legislation.

Both in Latvia and Lithuania, the shift of decision-making powers from the ministry to the subordinate bodies has been a clear tendency – in Lithuania, to the regional departments of the ministry, and in Latvia, partly to the regional boards and to the central agencies subordinated to the ministry.

The structures of environmental administration in Latvia and Lithuania described above have a clearly vertical orientation, whereas in Estonia the environmental administration remained for much longer horizontally oriented in its structure. The regional environmental protection authorities were part of the county administration, which was at first financed by local government budgets. The administrative reform took place in 1993, when the two-tier local government system was changed into a one-tier system, and the counties became directly subordinated to the central government. The regional environmental administration became part of the newly established counties. Later, at the beginning of 2000, environmental authorities were separated from counties and reformed into regional environmental authorities under the Ministry of the Environment. Of the environmental tasks, only supervision of land-use planning remained the responsibility of the counties (Peterson, 2004).

In 1994, the administration in Lithuania was restructured. Ten counties were set up as subordinate bodies to central government in order to carry out the regional policies of the state government (Ruutu and Johansson, 1998, p67). However, this did not influence environmental administration as the regional departments of the Ministry of Environmental Protection stayed separate from the counties. A later reform in 1997, in which the Ministry of Urban Development and Construction was merged with the Ministry of Environmental Protection, transferred some environment-related tasks of the counties, such as spatial planning, to the Ministry of Environment.

It seems that a common feature of environmental administration in all Baltic states is that environmental administration tends to develop separately from the counties, which serve as regional bodies of the central government for most issues. Discussions have from time to time considered whether the regional environmental administration should be merged with the county organization. The main argument against the merger has been that subordinating environmental decision-making under the governor's power would lead to conflict of interest since the governor is also responsible for decisions on developmental goals.

On the other hand, the integration of environmental issues within other sectors will require much more cooperation between the regional authorities in the current situation of separate administrative bodies. Another trend seems to be that decision-making powers are shifted, to a growing extent, to the regional bodies, and the ministries focus on policy-making and EU policy.

Local authorities

When the Baltic states restored their independence, they adopted the local government systems that existed during the first period of independence, in the 1920s and the 1930s. This was a two-tier system of municipalities and counties with elected councils. After a few years, Estonia and Lithuania streamlined this into a one-tier system, which was thought more suitable for a small country. Latvia, however, retained the two-tier structure (Ruutu and Johansson, 1998, p57). In this reform, Estonia based the new municipalities on former local governments and transformed the counties into state administration units. Lithuania chose to reform the former counties into municipalities. The number of municipalities turned out very differently: 56 in Lithuania and 254 in Estonia (Ruutu and Johansson, 1998, pp60, 68). In Latvia the number of local municipalities and self-governmental districts is 563 and 26, respectively. Seven of the local municipalities (i.e. the biggest cities) have the functions of both local municipalities and districts.

This chosen structure has led to quite a remarkable difference in municipality size. In Lithuania, the average population of a municipality amounts to 66,000 inhabitants; in Latvia, the average figure is 1700; and one third of municipalities have fewer than 1000 inhabitants. In Estonia, the average population in towns amounts to 22,500, and in rural municipalities, 2300 (Ruutu and Johanssen, 1998, pp60, 63, 68). Evidently, the size of the municipalities affects their resources and their means to perform tasks.

Another factor influencing the availability of resources is the limited rights of municipalities to levy taxes. At the moment, they are dependent upon state funding. However, there are continuous arguments on how to divide the revenue from income tax between the state and the municipalities, and also between rich and poor areas. Dependency upon the state budget limits the opportunities for municipalities to make long-term plans for their future activities.

The division of tasks between state authorities and local authorities changed during the 1990s. Local land-use planning (urban plans and detailed plans) has clearly become a municipal task, whereas the higher tiers of planning have remained in the hands of state authorities. Legal issues of environmental protection (i.e. implementing environmental legislation, issuing permits, conducting EIAs and inspection) have been the task of state authorities. Environmental services such as waste and water management have become the tasks of municipalities. In Latvia and Estonia, it has been noted that most of the municipalities were too small for these tasks and regional organizations were established. This trend is caused by pressures from EU directives, since management of landfills, for example, is strongly regulated and therefore larger units are needed. This has lead to the establishment of regional landfills, and the most common model is to establish a joint stock company, owned jointly by the municipalities.

Since the municipalities are omitted from legal environmental regulation, many of the bigger municipalities have directed their environmental activities towards voluntary projects, such as Agenda 21, municipal environmental audits and so on. The advantage of this line of action is that for such projects it is possible to obtain funding from international sources. Moreover, the public relations value of the projects is high. In Lithuania, for example, the municipalities are encouraged to prepare voluntary environmental programmes.

Environmental administration in the Baltic states has evolved substantially in just over a decade. The model that they seem to be aiming towards resembles the Nordic one. However, there are still some areas where development remains embryonic. A greater role for local authorities requires increased resources and strengthened capacities.

The style of governance has also been changing from an expert-centred administrative style to involving stakeholders and civil society in environmental policy-making. At the moment, the individual citizen has numerous opportunities for participation because of the new procedures brought along by environmental legislation – EIAs, strategic environmental assessments (SEAs), land-use planning and permits – and the Århus convention. But so far, only a few interest groups have developed the lobbying organizations necessary to enable participation in policy-making.

Spatial planning in connection with environmental impact assessments

Spatial planning

EU involvement in spatial planning has been minimal, and no directives stipulate how to arrange it nationally. The EU launched the European Spatial Development Perspective (ESDP) in the 1990s, which defines policy objectives at the union level and general principles of spatial development (European Commission, 1999). The ESDP is not based upon a codified process. Indeed, the ministers responsible for spatial planning decided that the ESDP would not have any normative value (European Commission, 1997). The development of spatial planning systems is left to member states. Therefore, spatial planning in the Baltic states was also grounded on national preferences.

The Baltic states originally approached spatial planning from a common base. During the Soviet period, the spatial planning regime was strong. Plans were prepared centrally in few governmental planning institutes, the process was non-democratic and the documents were secret. Furthermore, plans were developed for urban areas only, market forces did not exist and culture, natural heritage or environmental considerations were not included (Kule, 1999). Despite the similar starting point, Baltic states chose completely different approaches towards spatial planning. For the few first years after independence, Latvia almost forgot that planning ever existed since the responsibility was not assigned to any ministry or central government body. In contrast, Lithuania established the Ministry of Urban Development to bear responsibility for planning. Estonia also had a slow start and introduced the legislation only in 1995, but quickly established a well-functioning planning system, facilitated by assistance from Finland and Sweden.

In 1994, the Latvian Cabinet of Ministers approved the Regulations on Physical Planning. The Ministry of Environmental Protection and Regional Development was given the responsibility over state and regional-level planning and the Riga agglomeration. District authorities were given responsibility over district planning, and local comprehensive and detailed planning was assigned to municipalities. Several bilateral projects between Latvia and Finland, Belgium and Denmark were initiated to train the staff at different levels, mainly through learning by doing (Kule, 1999).

Despite the efforts by both the state and donors, many of the districts and most of the municipalities in Latvia did not take action to implement planning regulations. No qualified staff were hired and no plans drafted despite the schedules set by the regulation.

In 1998, the new planning regulations were approved. The responsibilities and approval procedure for planning were defined more clearly, and the responsibility for planning in Riga was transferred to Riga district. Municipalities and districts were given the right to freely agree on planning. The interests of the state were secured by state-level plans and district plans. Regional plans and local master plans were voluntary, and detailed plans could even be financed by the private sector (Kule, 1999).

All in all, the Latvian planning system was very liberal, and there was no attempt to force regions and municipalities to carry out planning. One reason for the slow start of planning practice was the lack of resources. Due to the large number of municipalities (see Chapter 8) and their small size, the financial basis was not very sound. Other tasks assigned to municipalities, such as education and healthcare, consumed most of the resources and only 0.8 per cent of budgets were used for environmental protection and land-use planning (Streips, 1994). In the end, spatial planning was not much used as an environmental policy tool to steer the location of activities.

In Lithuania, the emphasis on planning was much greater from the beginning of the 1990s. Lithuania had deep-rooted traditions of spatial planning since the spatial planning expertise for the Baltic States was concentrated in three Lithuanian planning institutions during the Soviet era (UN, 1999). The Ministry of Construction and Urban Planning was established immediately following the restoration of independence.

In 1993, the Temporary Rules of Planning of the Territories were adopted. Preparation of the Territorial Planning Law began in 1994. The law was adopted by the *Seimas* on 12 December 1995 and enforced on 1 January 1996; it is supported by subsequent rules and regulations. According to the Act, there are four levels of planning: national and regional levels under responsibility of state authorities; local municipalities; natural entities; and legal entities. The territorial plans are divided into three categories: comprehensive, special and detailed plans. The law provides detailed regulations on comprehensive and detailed planning (VASAB, 2000).

Municipal reform during the early 1990s resulted in 56 local authorities in Lithuania, reasonable in size to secure resources. Unlike other Baltic countries, Lithuania does not suffer from proliferation of small local authorities (World Bank, 2002). In 1999, the *Seimas* adopted an amendment to the Law on the Territorial Administrative Units and the number of municipalities was raised to 60.

One problem limiting the ability of municipalities to make long-term plans and perform their tasks is the conflict between the central and municipal governments over the division of revenues and expenditure responsibilities (World Bank, 2002). However, currently, the municipalities have a reasonable resource base to perform the planning tasks. The ten county-level administrations in Lithuania are essentially administrative units of central government (World Bank, 2002) and are responsible for regional planning.

Spatial planning in Estonia is regulated by the Planning and Building Act that came into force on 22 July 1995. Land-use planning is a prerogative of Estonian municipalities. Local development and building is guided by comprehensive and detailed plans, the latter being the basis of short-term building activities (Blechschmidt, 2003).

Environmental impact assessments

The ecological expertise procedure was the predecessor of EIAs in the Baltic countries. It was based on an environmental policy tool (State Environmental Expert Review), introduced during the last days of the Soviet Union (Cherp, 1999).

All of the economic activities had to go through an expert examination to determine if the project posed any threats to the environment. By nature, the procedure was expert oriented and did not include public participation. The evaluation focused on technology, unlike the Western EIA, which highlights impacts and leaves the choice of technology up to the developer. The ecological expertise procedure was performed in one step, whereas the EIA is a process-oriented evaluation. The result of the ecological expertise procedure was a binding permit to proceed with the development in contrast to an EIA, which is often mere recommendation to be taken into consideration in a separate permit decision.

In all Baltic states, the ecological expertise procedure was interlinked with land-use planning. In Estonia (Lass, 1999) and Latvia (Ernsteins and Benders, 1999), it was stipulated that each land-use plan should go through an ecological expertise procedure before being adopted. In Lithuania, territorial planning regulations included a type of EIA in the form of Initial Assessment of Planning Solutions. In Estonia and Latvia, the role of ecological expertise procedures resembled an *ex-post* quality control of the land-use plan, whereas in Lithuania the approach was more precautionary: the initial planning ideas were subject to assessment.

Due to the weak enforcement of land-use planning regulations in Latvia, the amount of land-use plans that were prepared during the 1990s can be counted on one hand. As a result, no experience in assessing land-use plans developed over time. Estonia's 1992 regulation also faced serious difficulties. Jalakas (1999) notes that there little experience with such evaluations until 1997. The expertise and environmental assessment of plans was abandoned in 2003 when land-use planning legislation was completely renewed and the assessment requirement was integrated within this new legislation. In Lithuania, the land-use planning system was stronger, and the evaluation of plans became routine.

EIAs are conducted in Baltic countries that developed in stages towards conformity with EU directives (337/85 and 11/97). Several attempts were made; but soon after adoption of the Acts a new round of preparation commenced. On the other hand, the EIA directive was also evolving and the amended directive was only adopted in 1997.

In Lithuania, the ecological expertise procedure was replaced by an Environmental Impact Assessment Act in 1996. This Act was based on the EU directive (337/85); but shortly after its adoption, a gap assessment was made that revealed several shortcomings. The Act was thus completely revised, rather than just amended (Kontio, 1999). Establishment of the Act encountered pressure from land-use planners and politicians. In their opinion, territorial planning legislation should supersede the EIA, as the territorial planning legislation already had some form of assessment. However, this approach posed serious difficulties when trying to transpose the requirements of the directive. More than 100 different draft versions of the EIA Act were presented to the executives of the ministry during a period of four years. The Act was finally approved in 2000, fulfilling the requirements of the EU directive (11/97) (Raulinaitis et al, 2001).

In Estonia, EIA legislation was included within the 1995 Act on Sustainable Development, and it was amended in 1997. These stipulations did not fulfil the requirements of the EIA directives and a new EIA Act was adopted in 2000 (Peterson, 2004).

In Latvia, the first attempt to develop a Soviet-style ecological expertise procedure was made in 1992. In 1996, a project was launched to develop the procedure to fulfil the requirements of EU Directive 337/85. The new Act, which was in conformity with EIA Directives 337/85 and 11/97, was adopted in 1998.

The main discussions concerning the EIA bills in Estonia and Lithuania centred on the overlapping of EIAs and land-use planning legislation. In Latvia, such discussion never emerged; on the contrary, the land-use planners understood EIAs as a new planning tool, hoping that they would also strengthen their sector.

The preparation of EIA legislation in Latvia and Lithuania was participatory. Several workshops were conducted for drafting the legal texts. Representatives of regional environmental authorities participated, along with representatives of bigger municipalities, environmental NGOs and industrial associations. This approach was quite new in the Baltic states, although in Latvia, a similar process had been implemented in preparation for the National Environmental Policy Plan. Before the mid 1990s, the preparation of legal Acts had been expert driven, and official comments on the final draft were asked from only a limited number of outside experts.

During the late 1990s, emerging industrial associations took a more active role in the preparation of new legislation. This was also the case with EIA legislation. However, in comparison with the Nordic countries, where environmental policy-making is corporatist and done in consultation with powerful industrial organizations, Baltic lobbying organizations were just taking their first steps.

As a result of the transposition efforts, all three Baltic countries adopted EIA Acts that were in conformity with EU Directives 337/85 and 11/97. In Latvia, such an Act was adopted in 1998, while Lithuania and Estonia adopted similar Acts in 2000.

The adopted EIA systems in all three countries closely resemble the Finnish and Norwegian systems, where the responsibility for conducting the assessment lies with the developer, and the environmental authority as a competent authority is responsible for ensuring that the assessment procedure is properly carried out.

Lithuania, with a fairly advanced land-use planning system, had the opportunity to adapt the Danish approach, where EIAs are interwoven into land-use planning systems, and the responsibility for assessment lies with planning authorities. On the contrary, in Latvia it was clear that the almost non-existing land-use planning could not serve as a framework for EIAs.

In the end, Latvia's and Lithuania's EIA systems turned out to be independent from land-use planning, featuring one competent authority (the Ministry of the Environment in Lithuania and the newly established EIA bureau in Latvia). The outcome was probably influenced by assistance projects: the Nordic–Baltic EIA project, led by Norway during 1993 to 1996, and the Finnish EIA projects on transposition of the EIA directives in Latvia (1996 to 1998) and Lithuania (1997 to 2001).

In Estonia, EIA also retained their independent status from spatial planning; nevertheless, EIA requirements were imposed on planning. However, this changed in 2003 when planning was released from the EIA requirement and environmental assessment paragraphs were incorporated within the planning law. In 2005, the situation changed again since implementation of the Strategic Environmental Assessment (SEA) Directive (42/2001) puts spatial planning in Estonia back under the environmental assessment requirements (Veis, 2004)

DISCUSSION

Environmental governance has been defined differently in varying contexts. According to Roderick Rhodes, the concept of governance is currently used in contemporary social sciences with at least six different meanings: the minimal state; corporate governance; new public management; good governance; social-cybernetic systems; and self-organized networks (Rhodes, 1996, p652). Lundqvist (1996, p15) defines governance as a question of policy content, organization, policy process and capacity of political systems to deal with environmental problems.

The Organisation for Economic Co-operation and Development (OECD) defines the concept of good governance much more narrowly and more technically. In its opinion, the key elements of good governance are technical and managerial competence; reliability; predictability and the rule of law; accountability; who is accountable for what; transparency; openness and availability of information; participation; the basic element of democracy; and the priorities of individuals, communities and private businesses (OECD, 1999, p62, based on OECD, 1998).

Weale et al (2000, p1) define environmental governance as an institutionalized system of rule-making and rules. According to them, the nature of EU countries' environmental governance is government without statehood; multilevel governance containing different tiers of authority; governance that is horizontally complex, featuring many actors at each level; and governance that is evolving and incomplete, raising questions of balance of authority, origin of policy, effective functioning and political significance (Weale et al, 2000, p6). Weale et al's concept originates mainly from EU and Western European experiences, and sees institutionalizing of environmental policy as organizational response to environmental issues. The complexity and cross-cutting nature of environmental policy-making is noted in this work (Weale et al, 2000, p193). This is not only limited to the national level; regional and local institutions are important as well.

When we consider these stated aspects of environmental governance in the context of the Baltic states, it seems that the focus in this region has been to modify the policy processes and administrative systems in a similar fashion to European countries. Progress has not been led by the need to deal with acute environmental problems or with an environmentally aware public. On the other hand, Kratovits (2001) maintains that despite the differences in importance of environmental problems in the Baltic states, they have first joined conventions with less stringent obligations, and the positive trends in the state of the environment have, rather, been influenced by local economic development than by the requirements dictated by international conventions.

It seems that the evolution of environmental governance in the Baltic states has, to a great extent, been state led. As described by Rinkevicius (2000), the Green movement lost its momentum after independence was re-established. However, some signs of multilevel governance can be seen, although these have been vague due to the lack of resources at the local level and the weak position of environmental NGOs.

Division of policy-making and administrative responsibilities depends upon the institutional context of the country, and the environmental administration cannot be looked at in isolation (Weale et al, 2000, pp200–201). In the Baltic countries, there has been an inevitable trend to decentralize governance. More and more, administrative functions are distributed to the regional and local levels. However, this requires constant strengthening of capacity and resources, and it has therefore taken more time to develop state-level governance.

According to Lundqvist (1996, p14), national structures and decision-making processes are dynamic; they develop over time and change in ways that may not be similar. This seems to be true for the Baltic countries where progress has been extremely dynamic. They have come with environmental governance through somewhat similar stages that can be identified in many Western European countries; however, the pace has been extremely fast. In the process, some stages have been ignored. Divergence in arranging the national structures has also been evident, wherever there is space for it. It is astonishing how quickly the Baltic countries have adopted modern European environmental policy tools and legislation, and reformed their administration. The pace of change, particularly with regard to transposition of EU laws where only cost-intensive measures were postponed, highlights the adaptability of environmental governance systems in the Baltic states. It seems that the political-administrative systems of the Baltic states are, at the moment, tuned up for rapid changes. This is highlighted by the fact that Baltic countries were among the first member states to transpose the requirements of the Directive on Environmental Assessment of Certain Plans and Programmes

(2001/42/EU). Old member states (e.g. Finland and Germany), on the contrary, were struggling with difficulties in transposing it as late as 2005. It seems that young organizations are dynamic and flexible.

The capacity of the Baltic states to adopt the new policies at a practical level is an entirely different question. The EU Commission has paid a lot of attention to the implementation deficit, and this may be a problem for new member states. Their economies are not as strong as those of the old member states despite speedier growth. Due to the constant capacity-building for the past ten years, their capabilities of truly implementing the requirements are growing. Some of the new policy tools have received a welcome response. Environmental impact assessments, integrated environmental permits and environmental management systems are all viewed as positive instruments in addressing environmental problems. Other policy instruments have only recently been introduced, and so it is too early to evaluate their performance. It should be noted that public authorities are not the only relevant actors in this field. The private sector also has an important role to play, as do financial institutions and the growing demands of consumers in environmentally conscious Western markets.

REFERENCES

Blechschmidt, K. (2003) 'Spatial planning in Poland and Estonia for promoting energy efficiency and renewable energy sources – theoretical Background', in *eceee 2003 Summer Study Proceedings*, www.eceee.org/library_links/proceedings/ 2001/

Cherp, O. (1999), 'Evaluating SEA systems in countries in transition', in *Baltic EIA, International Conference on the Environmental Impact Assessment, 28–29 April 1999*, Pärnu, Estonia

Compendium of Spatial Planning Systems in the Baltic Sea Region Countries (2000) VASAB, http://vasab.leontief.net/introduction.htm

Encyclopedia.com (2002) www.encyclopedia.com/Default.aspx, accessed 7 September 2005

Environmental Protection Ministry of Lithuania (1995) *Lithuania's Environment: Status, Processes, Trends*, Environmental Protection Ministry of Lithuania, Lithuania

Ernsteins, R. and Benders, J. (1999) 'Developments of understanding and preconditions towards strategic EIA in Latvia', in *Baltic EIA, International Conference on the Environmental Impact Assessment, 28–29 April 1999*, Pärnu, Estonia

Estonian Government (2005) *Official State Web Centre*, www.riik.ee/en/, accessed 20 January 2005

Estonian Institute for Sustainable Development (2001) *Proceedings of the Fourth Conference on the Environmental Conventions and the Baltic States: Facing the New Millennium*, Estonian Institute for Sustainable Development, Tallinn

Estonian Ministry of Environment (2002) *Estonian National Report on Sustainable Development 2002*, www.envir.ee/eng/sustainable.html#3, accessed 12 December 2004

Estonian Ministry of Environment (2005) *Estonian National Strategy on Sustainable Development, Sustainable Estonia 21*, www.envir.ee/orb.aw/class=file/action=preview/id=166311/SE21_eng_web.pdf, accessed 2 February 2007

European Commission (1997) *Report on Community Policies and Spatial Planning*, Working document of the EU Commission Services, Brussels

European Bank for Reconstruction and Development (1995) 'Environments in transition', *The Environmental Bulletin of the EBRD*, spring, EBRD, London

European Commission (1999) *ESDP European Spatial Development Perspective, Towards Balanced and Sustainable Development of the Territory of the European Union, 1999*. Agreed at the Informal Council of Ministers responsible for Spatial Planning in Potsdam, May 1999, European Commission, Brussels

European Parliament (1998) *Environmental Policy in Estonia*, Directorate-General for Research, Division of the Environment, Energy, Research and Stoa, Briefing no 5, European Parliament, Luxembourg

Finnish Ministry of the Environment (2004) *Ympäristöyhteistyö rakentaa hyvää naapuruutta. Suomen ja Viron ympäristöyhteistyö 1991–2004* [*Environmental Cooperation Builds Up Good Neighbour Relations: Environmental Cooperation between Finland and Estonia 1991–2004*], Finnish Ministry of the Environment, Helsinki

Jalakas, L. (1999) 'Strategic environmental assessment in Estonia; procedure and implementation', in *Baltic EIA, International Conference on the Environmental Impact Assessment, 28–29 April 1999*, Pärnu, Estonia

Kontio, P. (1999) 'Comparison of Lithuanian, Latvian and Finnish EIA Procedures', in *Baltic EIA, International Conference on the Environmental Impact Assessment, 28–29 April 1999*, Pärnu, Estonia

Kontio, P. and Karlin, A. (1999) *Latvian ympäristökysymykset ja koulutustarveanalyysi* [*Environmental Questions and Evaluation of Training Needs in Latvia*], Dipoli Series 3/1999, FiLEE Project (Interreg II), Helsinki University of Technology, Training Centre Dipoli, Espoo

Kratovits, A. (2001) 'The Baltic states and environmental conventions 1991–2000: general observations', in *Proceedings of the Fourth Conference on the Environmental Conventions and the Baltic States, Facing the New Millennium, 25–26 October 2001*, Haapsalu, Estonia

Kule, L. (1999), 'New environmental impact assessment legislation and spatial planning practice in Latvia', in *Baltic EIA, International Conference on the Environmental Impact Assessment, 28–29 April 1999*, Pärnu, Estonia

Lagerspetz, M. and Maier, K. (2002) 'Das politische System Estlands', in W. Ismayr (ed) *Die politischen Systeme Osteuropas*, Leske und Budrich, Opladen, pp 69–107

Lass, J. (1999) 'Environmental impact assessment in planning in Estonia', in *Baltic EIA, International Conference on the Environmental Impact Assessment, 28 – 29 April 1999*, Pärnu, Estonia

Lithuanian Ministry of Environment (2000) Ministry of Environment of the Republic of Lithuania, 2000

Lithuanian Ministry of Environment (2002) *National Report on Sustainable Development, 2002*

Lundqvist, L. (1996) 'Sweden', in P. M. Christiansen (ed) *Governing the Environment: Politics, Policy and Organization in the Nordic Countries*, Nordic Council of Ministers, Nord, Århus

Michanek, G. (1998) *Harmonisation of the Environmental Legislation in Latvia*, Ministry of Environmental Protection and Regional Development and Swedish Environmental Protection Agency, Latvia

Michanek, G. and Blumberga, U. (1998) *Environmental Legal System in Latvia*, Ministry of Environmental Protection and Regional Development and Swedish Environmental Protection Agency, Latvia

Ministry of Environmental Protection and Regional Development (1998) *Republic of Latvia: Overview*, Special edition of the Journal *Vide un Laiks*, Vides Projekti

National Baltic Sea Protection Programme (1996) Ministry of Environmental Protection and Regional Development, Latvia

National Environmental Action Programme, Latvia (1996) Ministry of Environmental Protection and Regional Development, Latvia

National Environmental Strategy, Estonia (1997) Approved by Parliament 12 March 1997, www.envir.ee/eng/strategy.html, accessed 3 January 2005

National Municipal Solid Waste Management Strategy – Programme 500 (1997) Ministry of Environmental Protection and Regional Development, Latvia

National Programme of Hazardous Waste Management (1996) Ministry of Environmental Protection and Regional Development, Latvia

National Report of Latvia to UNCED, 1992

National Spatial Plan ESTONIA 2010 (English summary) www.envir.ee/planeeringud/e2010en.html, accessed 3 January 2005

NEPP (*National Environmental Policy Plan for Latvia*) (1995) Ministry of Environmental Protection and Regional Development, Latvia

Nordic Project Fund (in cooperation with PI-Consulting Ltd and PI-EST) (1989) *Study of Environmental Protection: Estonia and Partly Latvia and Lithuania*, Nordic Project Fund, Helsinki

OECD (Organisation for Economic Co-operation and Development (1999) *Environment in the Transition to a Market Economy: Progress in Central and Eastern Europe and the New Independent States*, OECD, Paris

Peterson, K. (2004) Interview 30 November 2004, Helsinki

Purhonen, J. and Kontio, P. (2000) *Liettuan ympäristökysymykset ja koulutustarveanalyysi* [*Environmental Questions and Evaluation of Training Needs in Lithuania*], Dipoli Series 1/2000, Ecolift Project (Interreg II), Helsinki University of Technology, Training Centre Dipoli, Espoo

Raulinaitis, M. (2005) Pers comm, 5 September 2005

Raulinaitis M., Auglys, V., Buciunaite, I., Laurutenaite, B., Revoldiene, R., Kontio, P. and Punkari, M. (2001) *Manual for Environmental Impact Assessment in Lithuania*, Ministry of the Environment of the Republic of Lithuania and Finnish Environment Institute, Vilnius

Rhodes, R. (1996) 'The new governance: Governing without government', *Political Studies*, vol 44, no 4, pp652–667

Rinkevicius, L. (2000) 'Ecological modernization as cultural politics; Transformations of civic environmental activism in Lithuania', in A. Mol and D. A. Sonnenfeld (eds) *Ecological Modernization Around the World: Perspectives and Critical Debates, Environmental Politics*, Special Issue, vol 9, no 1, spring, pp171–202

Ruutu, K. and Johansson, M. (1998) *Division of Power and Foundation of Local Self-Government in Northwest Russia and the Baltic States*, Nordregio WP 5, www.nordregio.se/Files/wp9805.pdf

Ruza, S. (2005) Pers comm, 7 September 2005

Streips, K.L. (1994) 'Basic information on local governments in Latvia', in *Local Governments in the CEE and CIS*, Institute for Local Government and Public Service, Budapest

UN (United Nations) (1999) *Environmental Performance Reviews (EPR), Lithuania*, United Nations, Economic Commission for Europe, Committee on Environmental Policy, Environmental Performance Reviews Series no 4, United Nations, New York/Geneva

UN (2001) *Environmental Performance Reviews: Estonia Second Review*, Economic Commission for Europe, Committee on Environmental Policy, Environmental Performance Reviews Series no 12, United Nations, New York/Geneva

Vaht, Ü. (2001) *Estonia: Country Reporting on the National Assessment and Consultation Process for WSSD*, 26–28 September 2001, Antalya, Turkey, www.sdnpbd.org/wssd/preparatory-process/nationallevel/nationalassessment/56.pdf, accessed 3 January 2005

VASAB (2000) *Compendium of Spatial Planning Systems in the Baltic Sea Region Countries (2000)*, VASAB, http://vasab.leontief.net/introduction.htm

Vers, V. (2004) Pers comm, 30 November 2004

Water Supply and Wastewater Treatment in Small- and Medium-Sized Towns of Latvia (800+) (1997) Ministry of Environmental Protection and Regional Development, Latvia

Weale A., Pridham, G., Cini, M., Konstadakopulos, D., Porter, M. and Flynn, B. (2000) *Environmental Governance in Europe: An Ever Closer Ecological Union*, Oxford University Press, New York

World Bank (2002) *Lithuania: Issues in Municipal Finance*, Report no 23716-LT, Infrastructure and Energy Sector Unit, Latvia, Lithuania and Poland Country Unit, Europe and Central Asia Region

Part III

New Structures of Governance in the Baltic Sea Region

6

Governance beyond the Nation State: Transnationalization and Europeanization of the Baltic Sea Region

Kristine Kern and Tina Löffelsend

INTRODUCTION

After the end of the Cold War, the Baltic Sea Region developed into a highly dynamic area of cross-border cooperation and transnational networking. This trend was reinforced by the imminent enlargement of the European Union (EU), which increased by ten new members in 2004. Since the Baltic Sea is now surrounded by EU member states (with the sole exception of Russia), European integration appears to offer a real chance to clean it up: it is still endangered by pollution. EU enlargement resulted in a fundamental change in the governance of the Baltic Sea Region, although the new member states had already followed the lead set by Brussels and EU policies in the pre-accession phase.

However, the region is still divided into two parts: while the Nordic countries and Germany are considered environmental pioneers (Andersen and Liefferink, 1997; Joas and Hermanson, 1999; Lafferty and Meadowcroft, 2000), Poland and the three Baltic republics still lag behind European standards and face serious environmental problems that cannot be solved in the short term (Andersson, 1999; Reents et al, 2002). Thus, close cooperation between the countries and new forms of governance are necessary for the clean-up of the Baltic Sea as a common good and for the sustainable development of the entire region.

The aim of this chapter is to analyse different types of governance beyond the nation state in the Baltic Sea Region. The following section discusses the limits of national governance and describes four different types of governance beyond the nation state:

1   international regimes, such as the Helsinki Convention for the Protection of the Baltic Sea;
2   transnational policy networks, such as Baltic 21, the world's first regional Agenda 21;
3   transnational networks, such as the Union of the Baltic Cities (UBC); and
4   the European Union, with approaches such as the Northern Dimension for the development of the Baltic Sea Region.

These new governance types represent two parallel trends: a development to-wards transnationalization and the Europeanization of the Baltic Sea Region. Relevant examples and case studies are presented in the subsequent sections. First, the different aspects of Baltic transnationalization are discussed. Second, the Europeanization of the Baltic Sea Region is analysed in section four. In the final section, some conclusions are drawn regarding the emerging forms of governance in the Baltic Sea Region and the relationship between transnationalization and Europeanization.

GOVERNANCE IN THE BALTIC SEA REGION

The limits of national governance

In the past, environmental policy in the Baltic Sea Region was centred at the level of national governance. However, the sustainable development of the region can only be guaranteed through a combination of national governance and new modes of governance that reach beyond the nation state. Because the protection of the Baltic Sea as a common good is at stake, and because policy approaches, as well as impacts, vary considerably from country to country, different types of co-operative policies must be developed and applied systematically. Developments in the Baltic Sea Region reflect a general trend towards a new definition of the sovereignty of the nation state and the increasing importance of international organizations and regimes, on the one hand, and transnational and sub-national actors, on the other (Varwick, 1998, p56). However, this is not to say that the nation state has, or will, become obsolete. Despite these trends involving the obvious diffusion of power, authority and legitimacy to other government levels and actors, the role of the nation state remains crucial.

However, the importance of national government and governance in the Baltic Sea Region has declined. It is now defined and executed in new modes and arrangements beyond the nation state. Generally speaking, such new governance arrangements involve the transfer of national authority in three directions: upwards, to the level of international and supranational institutions; sideways, to civil society actors; and downwards, to sub-national actors (Rosenau, 1999, p293; Pierre and Peters, 2000, p83ff; Voelzkow, 2000, p281ff; Rosenau 2003).

What forms do these concurrent trends take in the Baltic Sea Region? First, responsibilities are already being increasingly reassigned to international and supra-national institutions. After the end of the Cold War, many new international and intergovernmental institutions such the Council of the Baltic Sea States were created and existing institutions such as the Helsinki Commission (HELCOM) gained momentum. These early 1990s developments were superseded by a comprehensive Europeanization of the Baltic Sea Region, which began in the mid 1990s. The European Union became the most important international actor in the region due to EU enlargement.

Second, many tasks that previously came under the authority of national governments were transferred from governmental to non-governmental actors. Such transfers continue and can be observed within nation states and at the international level where transnational policy networks (Benner et al, 2003) and public–private partnerships have emerged in recent years.

Third, since the early 1990s, government reforms have played an important role in the Baltic Sea Region, especially among the former socialist countries. Local and regional self-governance was reinstated in Poland, the three Baltic Republics and Russia. Decentralization and the devolution of authority have strengthened the position of sub-national entities and increased local capacities in these countries (Reents et al, 2002; Dorsch, 2003).

In sum, therefore, there appear to be two clear and strong tendencies towards the:

1   transnationalization of the Baltic Sea Region, because tasks are transferred from governmental to non-governmental and sub-national actors; and
2   Europeanization of the Baltic Sea Region.

Types of governance beyond the nation state

The emergence of various new forms of international, intergovernmental, supra-national and transnational governance in the Baltic Sea Region (Jann, 1993) was triggered by three developments that involved all levels of government and a wide range of policy areas:

1   the end of the Cold War, which led to the establishment of new transnational and sub-national actors, followed by the transformation processes in the former socialist countries, which were aided primarily by the Nordic countries;
2   the United Nations Conference on Environment and Development (UNCED) in Rio de Janeiro in 1992, which triggered the implementation of Agenda 21 (UN, 1992, p3) and introduced more integrative and participatory approaches;

3   increasing European integration, the product of two waves of enlargement in 1995 (Sweden and Finland) and 2004 (Poland, Lithuania, Latvia and Estonia), leading to the Europeanization of the entire region.

The Helsinki Convention, an international regime which had been in place long before the end of the Cold War, was complemented by transnational (policy) networks and superseded by the Europeanization of the whole region.

Governance by international regimes and intergovernmental cooperation

Governance by international regimes and intergovernmental cooperation, the first type of governance discussed here, is the traditional form of governance beyond the nation state. Traditionally, cross-border environmental cooperation is governed by international regimes and institutions agreed upon and coordinated by nation states. International regimes and intergovernmental cooperation in the area of environmental policy have played an important role since the 1970s (List, 1997; Young, 1999; Axelrod et al, 2005; Chasek et al, 2006). Nation states are still the dominant actors; non-governmental organizations (NGOs) and sub-national actors are not directly involved in decision-making. During recent years, however, such organizations have obtained observer status, although decision-making is still restricted to representatives of nation states (Oberthür et al, 2002). This classic model presupposes a strong nation state capable of implementing international agreements at sub-national level. The success of this model depends upon national capacities to induce changes at sub-national level in order to solve existing environmental problems.

Even during the Cold War period, cooperation between nation states across the Baltic Sea was comparatively close, particularly in the area of environmental policy (Bruch, 1999, pp68–69). The Helsinki Convention on the Protection of the Marine Environment of the Baltic Sea Area, signed in Helsinki in 1974, is an excellent example of this cooperation (Kindler and Lintner, 1993; Poutanen and Melvasalo, 1995). Nevertheless, until the end of the 1980s, the situation in the area was clearly dominated by national governance and the repercussions of the Cold War. After the end of the Cold War, intergovernmental cooperation increased rapidly. Examples of intergovernmental cooperation include the Council of Baltic Sea States (CBSS) (Stalvant, 1999; Hubel and Gänzle, 2002), founded in 1992, which aims to strengthen cooperation and coordination between the countries in the region; and Vision and Strategies around the Baltic 2010 (VASAB 2010) (VASAB, 2001; Görmar, 1997), an intergovernmental programme of the countries in the Baltic Sea Region in the area of spatial planning and development.

Governance by transnational policy networks in the Baltic Sea Region

Governance by transnational policy networks, a new type of governance, has emerged in recent years. Such policy networks incorporate actors from government, business associations, NGOs and local government networks. In contrast to the first type of governance discussed above, such networks are characterized by the integration of stakeholders in policy-making. Governmental, non-governmental and sub-national actors play similar roles within such transnational policy networks because all actor groups are involved in decision-making and policy implementation. The mode of governance changes because this kind of self-organization encompasses different actor groups. The change in decision-making results in the adoption of a different implementation model. Implementation is not restricted to national governmental regulation, but depends upon the initiatives of sub-national and non-governmental actors.

During the early 1990s, the massive environmental problems in the former socialist countries came to light and gained prominence on the regional political agenda. Therefore, major regional cooperation efforts focused on environmental policy and sustainable development. International and intergovernmental co-operation increased rapidly during this period. In 1996, four years after the Rio summit, the Council of Baltic Sea States finally initiated an integrative and participatory Agenda 21 process for the whole region (Baltic 21), which encompasses various policy sectors (such as agriculture or transport) and involves a large number of stakeholders.

Governance by transnational networks in the Baltic Sea Region

The emergence of governance by transnational networks, another new form of governance, is not directly related to international cooperation between nation states since it is a form of governance excluding (national) governmental actors. In the case of transnational networks, decision-making takes place in network organizations of non-governmental and sub-national actors. This mode of governance can be characterized as the self-organization of such actors. The implementation of internal decisions among civil society actors or networks of local authorities depends upon internal network governance. Since hierarchy cannot be adopted as an internal mode of governance, new governance instruments, such as benchmarking, have been developed and applied.

The emergence of numerous transnational networks in the Baltic Sea Region after the end of the Cold War is striking. Development in this region appears to be more dynamic than in other parts of Europe. A variety of transnational civil society organisations (e.g. the Coalition Clean Baltic, Social Hansa, etc.), economic organizations (e.g. the Baltic Sea Chamber of Commerce Association) or sub-national organizations (e.g. the Union of the Baltic Cities) were founded after the end of the Cold War, were developed very successfully and have begun to influence other governance types in the Baltic Sea Region. Self-governance by transnational networks developed as one of the new modes of governance and prompted a fundamental change of governance in the region.

Table 6.1 _Types of governance beyond the nation state in the Baltic Sea Region_

Governance by	Examples	Actors and modes of governance
International regimes Intergovernmental cooperation	Helsinki Convention Council of Baltic Sea States	Governmental actors Self-organization of nation states Hierarchical implementation strategies within nation states
Transnational policy networks	Baltic 21	Governmental, non-governmental and sub-national actors Self-organization of different actor groups Participatory implementation
Transnational networks	Union of the Baltic Cities Coalition Clean Baltic	Non-governmental and sub-national actors Self-organization of non-governmental and sub-national actors Implementation by internal network governance
Supranational institutions	European Union	EU as actor European multilevel governance Implementation by nation states and the recipients of EU funding

Governance by the European Union

European integration replaced national governance with multilevel governance causing political actors to interact across the different levels of government. This development encompasses the establishment of direct relations between the European Union and networks of local and regional actors, such as the Union of the Baltic Cities, a phenomenon typical of EU multi-level governance (Bache and Flinders, 2005).

The European Union has increasingly become a very important actor in the Baltic Sea Region. The Europeanization of the region gained momentum after the end of the Cold War, with the third EU enlargement in 1995, when Sweden and Finland joined. Another wave of integration and convergence began in the region when the European Union agreed to its fourth and latest enlargement. EU influence increased during the pre-accession phase of the fourth enlargement, which ended in May 2004 when the three Baltic Republics and Poland became full EU member states. So, within the last 10 to 15 years, the situation in the Baltic Sea Region has changed completely; during the late 1980s, the EU had no presence in the region because Denmark and West Germany were the only EU members at that time.

The Baltic Sea can be considered as a link between old and new EU member states. Today, the governance of the Baltic Sea Region is becoming more and more embedded within European multilevel governance. It is already evident that most governmental and non-governmental actors in the Baltic Sea Region orient themselves towards Brussels. This situation supports the convergence of all national initiatives around the Baltic Sea towards sustainable development. The EU developed its own approaches and policy instruments for the Baltic Sea Region – for instance, funding programmes such as INTERREG or TACIS. Today, EU regulations and EU funding shape the socio-economic and political development of the whole region.

THE TRANSNATIONALIZATION OF THE BALTIC SEA REGION

Governance by international regimes: The Helsinki Convention

In the countries neighbouring the semi-enclosed Baltic Sea, the need for co-operative initiatives to tackle regional problems was evident from an early stage. The process for establishing a common framework for environmental protection in the Baltic Sea Region started with two conferences in Visby, Sweden, in 1969 and 1970. However, the prevailing political climate was ill suited to the establishment of an international agreement between countries from the two antagonistic blocs. Closer regional cooperation in this area only became possible after the *rapprochement* of the two Germanys.[1] Thus, Finland offered to host a conference for the protection of the Baltic Sea at the Stockholm Conference on the Human Environment in 1972. One year later, in May 1973, a multinational expert meeting convened in preparation for the Diplomatic Conference on the Protection of the Marine Environment of the Baltic Sea in Helsinki. In March 1974, the Convention on the Protection of the Marine Environment of the Baltic Sea Area – commonly referred to as the Helsinki Convention – was signed by seven states (Fitzmaurice, 1992, pp47–50).[2] The ratification process ended in 1980 and the convention came into force in May of that year. In the context of the convention, it was decided that a governing body would be created, so the Helsinki Commission (HELCOM) was established (Bruch, 1999, p73). In 1992, the convention was revised, updated (e.g. concerning the list of harmful substances) and broadened in scope (e.g. now also encompassing inland waters, coastal zone management and biodiversity), making it more appropriate to the new political situation in the Baltic Sea Region after the fall of the Iron Curtain. The new convention was signed by all of the nine states that border the Baltic Sea and by the European Community.[3] It is a legally binding international treaty.

The Helsinki Convention is exemplary in character. During the Cold War era, it stood out as a shining example of East–West cooperation and as a symbolic exercise in peaceful co-existence. Furthermore, apart from its geopolitical significance, it was the first framework convention to encompass all aspects of the maritime environment and its protection, and it remains outstanding in its scope today (Bruch, 1999, p159). The Helsinki Convention regulates pollution 'from land or coast, waterborne or airborne, originating from the operations of ships, from pleasure craft, from sea bed activities, or from any other heterogeneous disposal at sea of wastes or other matters' (Fitzmaurice, 1992, p53). In terms of general principles, it incorporates the precautionary and the polluter pays principles, and promotes best environmental practices and best available technologies, environmental monitoring (of emissions) and the avoidance of risks (to health and the environment). In addition, the contracting parties commit themselves to implementing national protection measures and to engaging in international cooperation. Thus, they individually or jointly take all appropriate legislative, administrative or other measures to prevent and eliminate pollution (responsibility principle).

The Helsinki Commission is responsible for the coordination of intergovernmental activities. It is supported by a secretariat in Helsinki. The commission meets annually and also holds occasional ministerial meetings. Chairmanship rotates every two years among the contracting parties in alphabetical order. HELCOM recommendations are adopted unanimously and, although not legally binding, they must be taken into account in national legislation and environmental programmes. The purely advisory nature of HELCOM's decisions has not proved an obstacle in the past. Instead, what has emerged is that national capacities (particularly financial resources) and political will are the decisive factors for implementing HELCOM resolutions (Bruch, 1999, p93). In addition to the signatories, 19 international

NGOs have observer status in HELCOM. These include Local Governments for Sustainability (ICLEI), the Union of the Baltic Cities (UBC), Coalition Clean Baltic (CCB), and the World Wide Fund for Nature (WWF).

There are five working groups responsible for the handling of particular problems or sources of pollution.[4] The commission and its working groups manage 17 different projects, ranging from the maintenance of databases on hazardous substances and the development of measures for the preservation of the sturgeon population, to reviewing the risks posed by oil spills. The Monitoring Group (HELCOM MONAS) and the Land-based Pollution Group (HELCOM LAND) publish a report on the overall pollution load of the Baltic Sea every five years.

Of particular interest is the Programme Implementation Task Force (PITF), responsible for the coordination of measures and activities in connection with the Baltic Sea Joint Comprehensive Environmental Action Programme (JCP). This programme sets the environmental management framework for the long-term restoration of the ecological balance in the Baltic Sea. The JPC was established (together with the revised Helsinki Convention) in 1992 as the implementing agent whose task it is to reduce the Baltic Sea's pollution load; the JCP's mandate will run for 20 years. In terms of measures prescribed by the JCP, the emphasis is on investment in environmentally friendly technologies and the elimination of so-called 'hot spots' in the region. The elimination of the 162 identified pollution sources (hot spots) is an issue of primary importance.

HELCOM PITF projects often require cost-intensive investments (technology and infrastructure), on the one hand, and the collaboration of stakeholders, on the other. Accordingly, the PITF consists not only of representatives from the contracting parties to the Helsinki Convention, but also of spokespersons from international financial institutions, and governmental and NGOs.

In 2000, PITF started to organize regional workshops, convening stakeholders (national, regional and local authorities, owners of hot spots, international financial institutions, and NGOs) for the purpose of developing plans of action and allocating financial resources to tackle pressing environmental problems in designated areas. Workshops were held in most of the Baltic Sea states up to 2002, and HELCOM attributes a generally positive cost–benefit ratio to the workshops (HELCOM, 2001, p14). Between 1992 and 2001, 26 municipal and industrial hot spots were eliminated mainly through closures or production cuts at industrial plants. This number increased to around 80 in the intervening period because further hot spots were eliminated.

The initiatives of the Helsinki Convention encompassing approximately 110 HELCOM recommendations since the early 1980s contributed substantially to the improvement of the maritime environment in the region. The main achievements can be seen in the reduction of emissions and hazardous substances by at least 50 per cent[5] (primarily due to the elimination of hot spots); the adoption of stricter regulations for industrial emissions; the enactment of new legislation for the prevention of pollution by maritime traffic; the implementation of measures to avert illegal oil spills; and the improvement of regional environmental monitoring and assessment. However, many sources of contamination remain that are still polluting the Baltic environment with nutrients and hazardous substances. Intensive agriculture, inadequate municipal and industrial wastewater treatment facilities, oil spills and industrial discharges still pose major threats to the Baltic environment.

HELCOM has extended its activities significantly over the decades. It received much praise for its solid cooperation in environmental matters; but its expansion in this area also prompted criticism from NGOs who blamed HELCOM for being 'too large, expensive, and slow to act' (VanDeveer, 1999, p13). Around 40 to 50 meetings per year, which often lasted for several days, consumed much of the participants' time and resources. This has been a serious hurdle to smaller NGOs' ability to participate, and it may explain why only larger international NGOs and other umbrella organizations are represented in HELCOM.

The most important obstacle to environmental policy implementation is probably the lack of capacity in the transition states. In these states a combination of public-sector performance deficits and a generally lower-level awareness of environmental problems place serious constraints on policy implementation. HELCOM recognizes that some progress has been made – for example, in the Baltic states; but fundamental problems remain to be resolved in Russia. Thus, to be effective, HELCOM must adopt a three-dimensional approach that focuses on capacity-building with respect to human resources (training of personnel), development of the public sector (reforming bureaucracies and institutions) and raising of public awareness (VanDeveer, 1999, p10). HELCOM has acknowledged the need for these priorities and is focusing its efforts in this direction.

In order to avoid unnecessary duplication of tasks, HELCOM has recently made efforts to harmonize and consolidate its regulations and activities with those of other international regimes and organizations. HELCOM is working to bring its recommendations into line with the Convention for the Protection of the Marine Environment of the North-East Atlantic (OSPAR) resolutions and recommendations.[6] In 2000, a joint Baltic 21/HELCOM working group was established to coordinate the often overlapping activities of both organizations. Finally, HELCOM has begun to try to harmonize its regulations with EU regulation – an aim that is becoming increasingly more significant.

HELCOM's enhanced cooperation with the non-governmental sector, particularly in project-based collaboration, is a sign of a shift in its policy: the intergovernmental level alone is no longer recognized as sufficient for the successful implementation of environmental policy. To create awareness, legitimacy and acceptance of decisions requires the participation of societal actors. HELCOM's gradual opening up to civil society actors and other stakeholders represents a response to this requirement and marks the advent of a relaxation, to some extent, in the strict hierarchical order of this international regime, which has earlier dealt exclusively with governmental actors.

Governance by transnational policy networks: Baltic 21 – an Agenda 21 for the Baltic Sea Region

Transnational policy networks have a different structure than international regimes insofar as such networks involve a variety of actors ranging from nation states to civil society. Policy networks can facilitate cooperation between different partners in a certain policy field (e.g. sustainable development) by promoting a common agenda. Equality among the participating stakeholders is a prerequisite and acknowledges the importance of all participating levels for the successful implementation of the policy agenda.

Baltic 21 was initiated as a result of the adoption of Agenda 21 at UNCED in Rio de Janeiro in 1992. Baltic 21 was an initiative of the Council of Baltic Sea States; the heads of state and governments of CBSS member countries agreed together with the EU on the development of a regional Agenda 21 for the Baltic Sea Region. The agenda process was officially launched in 1996. A wide range of actors was involved in drafting sectoral reports as part of the development of the agenda's programme. These reports constituted the background for the Baltic Agenda 21, which was adopted by CBSS foreign ministers two years later (see Baltic 21, 1998, p4). The agenda initiative involves all countries surrounding the Baltic Sea (plus Norway and Iceland). The process is supported by a small secretariat in Stockholm that operates as a unit of the CBSS Secretariat. It has three staff members, as well as a consultant for specific priority projects. Procedural steering is carried out through the Senior Officials Group (SOG) that comprises some 40 parties, including representatives from national ministries, the European Commission (DG Environment), intergovernmental organizations (e.g. HELCOM, the International Baltic Sea Fishery Commission or Vision and Strategies Around the Baltic 2010), international financial institutions, (e.g. the European Bank for Reconstruction and Development or the World Bank), international sub-state and city networks (e.g. the UBC or ICLEI), international business networks (e.g. World Business Council for Sustainable Development or International Chamber of Commerce), international academic networks (Baltic University Programme), and international environmental NGOs (e.g. Coalition Clean Baltic and the WWF).

The wide range of representatives in the SOG reflects one of the principles adopted by the agenda process – namely, to be inclusive and open. Despite this aim, however, civil society representatives are largely outnumbered by governmental and other institutional actors. Furthermore, the SOG presidency rotates (every two years) only between countries and the EU. The SOG plenum elects a chairperson to head, and representatives from four countries to operate, its bureau, whose task it is to assist the SOG chairperson and secretariat. At present, the members of the bureau are all government representatives from ministries.

Baltic 21 strives to assist the countries of the Baltic Sea Region in their efforts to achieve sustainable development; but primary responsibility lies with the states themselves, who must ensure that the Baltic 21 goals are streamlined in accordance with national policies (Baltic 21, 2003, p8). Baltic 21 is designed as a long-term project for sustainable regional development. The time frame for Baltic 21 is five years; but the entire process is projected for a duration of 30 years, and it will go beyond environmental protection, encompassing the economic and social spheres as well. Seven sectors of the economy – namely, agriculture, energy, fisheries, forests, industry, tourism and transport – and, since 2000, education, have been identified as priority areas for planned collaborative activities in accordance with the Baltic 21 Action Programme. Cooperation was also established in the area of spatial planning. The Baltic 21 Action Programme (which complements the strategic part of Baltic 21 and which was adopted at the same time) lists 30 potential activities, primarily related to structural development and enhanced regional cooperation.

For each sector or area, priority indicators were established, goals and time frames set, and plans of action developed. The supervision of each sector or area is the responsibility of one or two SOG members, so-called 'lead parties'. All of the lead parties are countries or intergovernmental bodies such as the International Baltic Sea Fishery Commission (IBSFC).

In June 2003, the periodic 2003 report *Baltic 21 Report 2000–2002: Towards Sustainable Development in the Baltic Sea Region*, was released, reviewing and evaluating the Baltic 21 activities up to that time. According to this report, measures taken in the various sectors include assessment studies, networking between responsible or affected regional actors, and the development of guidelines or concepts (e.g. for fisheries or forestry). Other sectors, such as industry and tourism, have concentrated on the promotion of environment-friendly production or certification schemes. All sectors have been active in trying to establish links between respective stakeholders and in promoting exchange (see Baltic 21, 2003, p11–20). The report specifically mentioned 11 real projects, which have been implemented so far under the auspices of Baltic 21; these projects deal with concrete problems in the industrial sector (nine projects), spatial planning (one project) and transport (one project). Funding for these activities has been provided mostly by national agencies and governments, EU structural funds (INTERREG IIIB, PHARE and LIFE), and the business community.

In addition to the sectoral activities, Baltic 21 is also involved in 'joint actions' (JAs), targeting cross-sector issues. Seven such actions were launched, each of which had one responsible actor who initiated and managed the common activities (see Baltic 21, 2003, pp1–2). The picture concerning the implementation of JAs is equally mixed. For example, JAs 3 and 5 (Demonstration Areas and Pilot Projects and Procurement of Technologies) had not gone beyond an initial screening process for the evaluation of potential for future actions, while JA 4 (City Cooperation and Sustainable Development Issues in Cities and Communities) was considered particularly successful. The joint initiative is coordinated by the Union of the Baltic Cities. Building on its ongoing work in the area of environmental cooperation at local level, UBC completed 11 co-operative projects between 2000 and 2002; these activities were carried out within the UBC's own Agenda 21 Action Programme, and involved three-quarters of its member cities. The projects encompassed different approaches to sustainable urban development – for instance, environmental management or best practice exchange. The 2003 Baltic 21 assessment report refers to this success and the UBC's contribution to the integration of Baltic sustainability initiatives within similar processes at European and international level. Consequently, Baltic 21 (2003, p23) demands that resources for activities be transferred from the national to local level.

The first *Baltic 21 Biennial Report* (2000) primarily noted that activities were still in an initial kick-off phase one-and-a-half years after the adoption of the agenda. The 2003 assessment report paints a mixed picture of developments. Generally, however, it is not over-enthusiastic about the achievements of the Baltic Agenda process. Acknowledging the general goodwill of all parties, it acknowledges differences between sectors: progress has been made in energy, fisheries, industry and spatial planning. Some progress was noted for forestry, and a 'good start' recognized for Baltic 21 activities in education; tourism, agriculture and transport, on the other hand, lag behind. Agriculture, tourism and fisheries were identified as notoriously difficult sectors where much action is still required. Thus, the fact that the 2003 report identifies two of these problematic sectors as belonging to the group of laggards in Baltic 21 must be interpreted as a special challenge. In terms of the main obstacles to the success of Baltic 21, the report identifies a lack of commitment in some of the responsible ministries (Baltic 21, 2003, pp4–5). Since the overall success of the agenda largely depends upon resources provided by nation states and measures implemented by them, this is a crucial factor. The report concludes that sufficient resources and organizational stability are what paved the way for progress in the successful sectors.

The influence of states and civil society in the Baltic 21 process is not evenly balanced. Nation state representatives hold the most influential posts and contribute the most funding. Whether by accident or design, this inevitably results in a significantly greater share of influence. This could be for practical reasons (e.g. capacity and legitimacy); nevertheless, it prevents other sub-national or civil society stakeholders from gaining more influence. Furthermore, the current leadership is dominated by a small number of older EU member states. Given that Baltic 21 claims to be an all-inclusive process, providing equal status to all participants, it must overcome this bias in order to become as democratic and open as stated in its agenda. The 2003 Baltic 21 report also addresses these issues and draws some substantial conclusions. It demands the general revision of Baltic 21's mandate, visions and indicators. The report acknowledges the inherent conflict of objectives between economic development and environmental goals, and identifies greater involvement in European and international sustainability processes as a solution.[7] A new strategy for the Baltic 21 process was proposed in spring 2004, with the vision to pursue sustainable development in the Baltic Sea Region by regional multi-stakeholder cooperation. Moreover, this proposal emphasizes the strengthening of cross-sectoral work and the development of a selected set of

Lighthouse Projects, which are designed to ensure high visibility and engage as many participating countries and sectors as possible in proving the added value of sustainable development (*Baltic 21 Newsletter* 1/2004, p5; Baltic 21 Press Release, 6 April 2004).

The institutional arrangements for Baltic 21 differ substantially from the traditional forms of international governance (international regimes and inter-governmental cooperation) because the agenda process is based on a concept of broad stakeholder participation, including governmental (plus the EU) as well as non-governmental actors. This is certainly due to its origin in Agenda 21, which also emphasizes the broad participation of non-governmental and sub-national actors. Baltic 21 has been especially successful in areas where local actors became directly involved, for example, in the area of city cooperation in the framework of Joint Action 4, which was carried out by the Union of the Baltic Cities.

Governance by transnational networks: The Union of the Baltic Cities

During recent years, many transnational networks have emerged in the Baltic Sea Region, giving rise to new forms of governance beyond the nation state. Cooperation is not restricted to civil society actors, but also encompasses collaboration between sub-national actors (i.e. networks of cities and regions) which are often neglected in this context. In contrast to 'governance by transnational policy networks', discussed above, 'governance by transnational networks' is not limited to the involvement of non-governmental or sub-national actors in decision-making or implementation. Emphasis is put on the fact that transnational networks 'govern' their members and that policy convergence among their members can be achieved through new modes of internal network governance, which do not require governmental actors for decision-making or implementation. Therefore, 'governance by transnational networks' is a form of private governance 'without the nation state'.

With the decrease in national sovereignty and growing limitations placed on national steering capacities, European municipalities' scope for action is increasing. This means that towns and cities have the opportunity to enter the European and international political arena and develop into global players. Their participation is often actively supported by international and supranational bodies such as the European Commission, thus creating new 'glocal' governance arrangements beyond the reach of nation states, consequently further undermining national authority. At the same time, policy-making in municipalities is affected by global change (e.g. environmental challenges) and by decisions taken at supranational and international level. Thus, the international and transnational involvement of municipalities is imperative. Networking and collective articulation of interests is more essential than ever for actors at local level to make their voices heard in European or international contexts. Consequently, a large number of such organizations have been established since the late 1980s (Kern, 2001; Kern and Bulkeley, 2008).[8] In line with similar international trends, advanced political integration (through the EU, in particular) obviously produces post-national structures, which facilitate and require collaboration at sub-national level and across borders. These networks are generally characterized by a horizontal, polycentric and non-hierarchical structure.

Cooperation among cities around the Baltic Sea displays specific features related to the longstanding tradition of their relationships (Vartiainen, 1998; Groth, 2001). In this case, too, the collapse of the Iron Curtain was the impetus for the revival of transnational relations. For example, the first free municipal elections in Poland in June 1990 provided Polish cities with new authority and legitimate self-governance. Thus, on the initiative of the mayors of Gdansk and Kalmar (in Sweden), a conference was held to establish an association of cities in the Baltic Sea Region. Then, in September 1991, the Union of the Baltic Cities was founded in Gdansk by 32 cities from 10 countries around the Baltic Sea. The UBC was set up as a general network, offering a platform and a 'tool' for the activities and interests of its members 'in a wide spectrum of spheres of interest'. The UBC considers itself primarily as an advocate of its members' interests and 'for the Baltic Sea Region as such' (Engström, 1998; Wohlgemuth, 1998; Kern, 2001; UBC, 2001a and b; Lindström and Grönholm, 2002).

The UBC's membership has more than tripled since its founding and, today, the organization has over 100 members, with almost 90 per cent of the founding members still participating. One possible major reason for this dynamic development is that the UBC follows the tradition of the old Hanseatic League and is therefore probably particularly attractive to many former Hanseatic cities. A remarkable fact about the UBC is that cities from older EU member states and those from the newer member countries which have just recently joined the EU (formerly 'transition states') are quite evenly represented. The fact that the UBC was launched as a Swedish–Polish initiative was probably an important determinant in this respect. Moreover, joining the UBC was principally motivated not by the specific characteristics of the municipalities involved (especially size), but rather by their sense of belonging to the Baltic Sea Region.[9] The UBC's organizational structure is highly developed. Its most important organs are the general conference (which meets biannually), the presidium (president and two vice presidents), the executive board, the secretariat and ten commissions. The UBC Secretariat is located in Gdansk and was financed by that city from the outset. The commissions cover the entire range of (transnational) urban policy:

- business cooperation;
- culture;
- education;
- environment;
- health and social affairs;

- information society;
- sports;
- tourism;
- transportation; and
- urban planning.

In addition, there are networks that fulfill cross-sectoral functions, among them the Local Agenda 21 Network and the Women's Network. An attempt is now being made to coordinate the activities of the commissions. To this end, regular meetings are held between commission chairs and the executive board.

Although the UBC is not a network specializing in environmental issues, the sustainability principle was incorporated as a goal in its statute and strategy.[10] Thus, sustainable development is considered as one of the UBC's key policies. The Commission on Environment (EnvCom) was one of the first commissions established by the UBC. EnvCom meets annually, and it is the central body for implementing strategies associated with sustainable development. In addition to this, the UBC also has a cross-sectoral Agenda 21 Working Group whose task it is to coordinate UBC member cities' Agenda 21 activities: the working group convenes two to three times a year. EnvCom and the Agenda 21 Working Group are both open to any and all interested UBC member cities. EnvCom and the Agenda 21 Working Group are administered through a secretariat in Turku, which, thanks to third-party project funding (in particular, from the EU and the Nordic Council), is better equipped than the UBC Secretariat in Gdansk.

In its initial phase, the UBC functioned mainly as an initiator of cooperation between the cities of the Baltic Sea. The primary task in this context was to establish contacts and create a basis for collaboration. The municipalities in the post-communist countries lacked basic equipment in fundamental sectors (schools, hospitals, transport, etc.) and their means of communication were often deficient; therefore, the need for assistance in these areas was pressing. The first five years of the UBC were thus characterized by concrete activities to secure direct aid. By the late 1990s, the situation of the municipalities in the transition countries had improved so much that the UBC could shift its focus to other policies and fields of cooperation. By then, the commissions were also becoming more differentiated and targeted in their work and new commissions – for example, for education and urban planning – were established (UBC, 2001b, p2).

Work on the UBC's Agenda 21 strategy also began during the late 1990s. EnvCom, working together with the Commission on Health and Social Affairs and the Women's Network, was highly involved in formulating the UBC Agenda 21 Action Programme. This programme was launched at the UBC General Conference in 1999, and was updated in 2001 and 2003. The programme contains 'policies, network service and project parts, as well as ... sustainability guidelines for member cities' (UBC, 2002, p6). Interestingly, 85 per cent of UBC member cities already pursue Local Agenda 21 activities on their own (Lindström and Grönholm, 2002; Joas, 2003). The UBC's Agenda 21 Action Programme is considered to be supplementary to those activities, providing service to UBC members and coordinating their various actions. Best practice exchange is high on the agenda and is promoted via workshops, seminars, twinning and the development of a set of European Common Indicators for Urban Sustainable Development as benchmarking tools. The Best City Practices Project (2000 to 2001) was relaunched and renamed Transferring Best Environmental Solutions between Towns and Cities (TBestC). TBestC's task is to pair up (twin) suitable cities for mutual policy learning in environmental protection. An award scheme, the Best Environmental Practice in the Baltic Cities Award, complements this approach to enhance benchmarking among member cities.

All of these activities indicate that the UBC is actively developing and implementing innovative measures for sustainable urban development in the participating cities. It is also active in representing its members in European and international policy arenas. This includes project-based cooperation with other networks such as Eurocities, as well as its involvement in the Baltic 21 process and HELCOM. In 2002, the UBC signed a cooperation agreement with Eurocities. Formalizing relations in this way is a response to the increasing need for better coordination among city networks. With the expansion of city networks, the streamlining of activities has become more important to avoid duplication and conflict — for instance, when competing for funds.

At the international level, the UBC was represented at the United Nations World Summit on Sustainable Development in Johannesburg in 2002. The UBC has also agreed to cooperate on a project with the Local Authorities of Lake Victoria Region in Africa. Internationally, however, special emphasis is placed on links with the European Union. The UBC promotes the appointment of EU coordinators in the municipal administrations to support UBC member cities in the European policy process because this kind of professionalism increases the chances for cities to obtain EU funding.

So far, the organization has been quite successful in attracting a balanced membership from all countries around the Baltic Sea. One problem with city networks, in general, and the UBC, in particular, is that they tend to attract municipalities which are already progressive, with the will and capacity to become involved in wider networks. Thus, laggard cities and towns may be left even further behind if they miss out on the opportunity to join such networks. A main concern for a regional network such as the UBC is to involve local authorities from both ends of the spectrum — the laggards as well as the pioneers. Inclusiveness is a prerequisite for sustainable regional development.

The UBC is, without a doubt, a good example of successful transnational (self-) governance beyond the nation state. Furthermore, it displays the novel features of such network organizations in maintaining transnational relations at the local level, combined with an active outreach towards the European level. The increasing Europeanization of the UBC becomes apparent upon closer examination of the

UBC Environmental Commission's budget, which reveals that most of its projects are funded from EU resources (UBC, 2004, p111–112). This is clear evidence for the Europeanization of this transnational network.

THE EUROPEANIZATION OF THE BALTIC SEA REGION

All governance types discussed show the increasing importance of European governance in the Baltic Sea Region. The European Union has become a central actor in environmental governance and sustainable development in this region. Although it may not be a dominant actor in all respects, the EU's position in decision-making has been strengthened; it plays a prominent role in creating frameworks and standards, targeting the achievement of policy integration, and promoting certain policies through targeted funding.

The role of the European Union in governance in the Baltic Sea Region is determined by the development of European environmental policy. During the early years, European environmental policy was based on a command-and-control approach, and executed via regulation (e.g. setting European standards and defining limits).[11] Over the past three decades, European environmental policy has evolved slowly into a 'context-based' and integrated policy concept involving new instruments, including the participation of a wider group of stakeholders. The former top-down process has become more diffuse, reflecting the multilayered nature of the European political system (Jordan, 2005; Lenschow 2005).

Four aspects relevant to this development are of particular interest in the context of this chapter. The first of these is 'subsidiarity', one of the EU's prime governance principles. This principle requires the EU to become active only if subordinated (national and sub-national) levels are not sufficiently equipped to respond to particular challenges or to implement certain policies. As a result, the incorporation of other (i.e. sub-national) levels of government within policy formulation and implementation has proven necessary for successful policy implementation.

The subsidiarity principle is closely related to the next point: the obvious implementation deficit in the area of European environmental policy. The debate about how to address this shortfall corresponds to similar ones at the national level. Despite all efforts, the EU has only been partially successful in implementing its environmental policy. A gap remains between administrative capacities and structural deficits in the member states (particularly in the case of the new members). This sometimes makes it difficult to put EU legislation into practice at local and regional levels. The more that institutional change is required for compliance with EU policies, the less likely effective policy implementation becomes. These problems can only be resolved if a wide range of stakeholders is included in the policy process. The success of environmental policy depends upon the 'positive mobilization ability' (Knill, 2003, p190) of affected and responsible actors. This insight opens the door to previously marginalized non-state actors from industry and civil society, and paves the way for new governance arrangements (Knill, 2003, p192ff).

Third, a new EU objective, policy integration (Jordan and Lenschow, 2000; Lafferty and Hovden, 2003) is highly relevant to the achievement of sustainable development in the Baltic Sea Region. With respect to the horizontal dimension, this strategy affects the integration of sectoral policies such as energy, transport and agriculture. However, policy integration also has a vertical dimension because it 'requires coordinated responses from *all* levels of government in the EU – European, national, regional *and* local' (Jordan and Lenschow, 2000, p111).

Fourth, in the White Paper on European Governance (European Commission, 2001), the EU outlined its aim to create more openness, participation, coherence and effectiveness. It called for the active involvement of local and regional authorities and for a more systematic dialogue with representatives from these levels through national and European associations. Thus, the commission intends to shape policies more flexibly and, hence, is responsive to 'regional and local conditions' (European Commission, 2001, p4). The commission proposed holding dialogues prior to formal decision-making. This concerns not only civil society actors but also associations of regional and local authorities.[17] Although the concepts and proposals of the European governance White Paper were certainly not new, for the first time they provided the EU with strategic governance guidelines for actual politics and policies. Practical relevance guidelines still need to be translated into the EU policy concepts in a coherent way.

All of these four factors are relevant to the governance for sustainable development in the Baltic Sea Region. In contrast to the Helsinki Convention's international environmental policy approach, which was developed 30 years ago, Baltic 21 as a sustainable development strategy covers seven economic sectors, as well as education and spatial planning. In addition, it encompasses joint actions targeted at local-level implementation. Thus, the trends that can be observed in the Baltic Sea Region correspond with the overall development of EU policy.

In addition to its direct involvement in policy-making, the EU was of particular importance to its new member countries in the pre-accession phase. These countries were required to adopt the *acquis communautaire* – that is, to draft and implement new legislation in compliance with EU standards and to shape their policies accordingly in order to fulfill the conditions for EU membership. These new requirements and measures created a foundation for environmental protection in the accession countries and throughout the entire Baltic region. Although the new member states still have a lot of catching up to do, the path is irreversible. By joining the EU policy-making machinery, the new member states are pushed forward. The role of the EU Commission, which has already changed and become more significant over the past several years, continued to evolve. In some countries, environmental awareness is lacking; there are only a small number of civil society groups concerned with environmental issues, and opposition to new environmental initiatives is strong. Hence, the positive mobilization required for successful environmental policy is often hindered by local conditions.

The EU also targets the special needs of the northern regions of Europe, particularly through the 'Northern Dimension'. This programme and its action plan are typical examples for the EU Commission's new policy strategy. The Northern Dimension was first established for the years 2000 to 2003, and renewed for the period from 2004 to 2006. It covers the EU's northern neighbours (Poland and the Baltic countries being new EU members) and focuses, in particular, on Russia.[13] Among other policy areas, it addresses and promotes cross-border cooperation and environmental policy, pursuing a multi-stakeholder approach.

The Northern Dimension relies on the willingness of partners; in its Second Action Plan, it calls for the inclusion of a wide range of stakeholders (including NGOs and local authorities) and sets soft targets (e.g. monitoring and reduction of environmental pollution by toxic substances). The Second Northern Dimension Action Plan 2004–2006 outlines projects and areas of action in the annex, but delegates the details and specific activities to other actors. For example, the plan calls for the implementation of the Baltic Sea Joint Comprehensive Environmental Action Programme and states that cooperation with Russia is the dominant strategic goal. The Northern Dimension itself will, in turn, be redefined in the context of the new neighbourhood policy: the 2003 Wider Europe strategy.

The EU has created a number of different regional policy instruments for projects in the Baltic Sea Region. These programmes include INTERREG, TACIS and PHARE, for cross-border cooperation, in particular, and LIFE for the environmental dimension. Each programme has unique responsibilities and specifically targets different groups of actors, countries and projects. EU financing of projects in the Baltic Sea Region is done primarily with EU structural funds. Under the Northern Dimension umbrella, international financial institutions set up the Northern Dimension Environmental Partnership (NDEP), which is supported by a special fund designated for environmental cooperation, particularly with Russia. Whereas the recipients are highly appreciative of these resources, anger is also widespread about the bureaucratic application procedures and the incompatibility of various instruments. HELCOM, Baltic 21 and UBC projects are funded through other EU programmes.

The three case studies have shown, therefore, that the European Union is directly involved in decision-making as a stakeholder: it is a signatory of HELCOM, it was involved in the initiation of Baltic 21 and, as an SOG member, it remains a partner in the Baltic 21 process. The European Union is generally responsive to the needs and challenges of sustainable development, on the one hand, and to European governance, on the other. It tries to address these goals by means of new governance arrangements and by assigning a greater role to non-state actors and, to a certain extent, by mainstreaming their involvement in the policy process. However, the soft policy approach taken by the EU on certain issues – for example, within the Northern Dimension, could also be questioned. Therefore, in some areas, a regulative approach promises better outcomes. Sustainable development needs a broad base of support at all levels of government. In this respect, the integrative approach of Baltic 21 seems to fit very well into the overall European strategy. The European approach of using a combination of 'old' and 'new' instruments can generally be considered on the right track, although it is not as successful as it could and should be. Despite the shortcomings, however, the EU is a relevant actor, an important partner and an influential promoter of sustainable development in the Baltic Sea Region.

CONCLUSIONS: TRANSNATIONALIZATION VERSUS EUROPEANIZATION OF THE BALTIC SEA REGION

Based on the analysis of different types of governance in the Baltic Sea Region, it is now possible to conclude that nation states are not obsolete and will continue to play an important role in the sustainable development of the region. However, national governance has its limits and new forms of governance beyond the nation state are crucial for the future development of the region. The general development can be characterized by both: the transnationalization and the Europeanization of the Baltic Sea Region.

With regard to the traditional mode of governance beyond the nation state – that is, international regimes – the most striking difference between the situation during the 1980s and the current situation is the strong position of the European Union within this international regime. The question remains as to the extent to which nation states are willing to grant stakeholders access to decision-making. HELCOM clearly shows that stakeholders are not fully integrated within policy formulation, which remains under the authority of the nation states (and the EU). This can be considered as typical for an international regime created 30 years ago. Even so, it is also obvious that HELCOM has been undergoing considerable change: governmental actors increasingly seek support from non-governmental actors for specific projects, particularly at local and regional level (e.g. the elimination of 'hot spots'). This means, at least, that non-governmental actors are gaining in significance in the area of regime implementation.

In contrast to HELCOM, Baltic 21 was designed as a multi-stakeholder network from the outset. The need for the broad participation of all relevant groups was recognized as essential for the success of the Baltic-type Agenda 21, which is directly related to Agenda 21 as adopted in Rio de Janeiro in 1992, and which stresses multi-stakeholder approaches for all Agenda 21 processes. Thus, the emergence of a new governance approach – that is, a transnational policy network – would seem to have been triggered primarily by international developments. The fact that Baltic 21 was the first regional Agenda 21 process can be explained by the unique situation that prevailed in the Baltic Sea Region at the end of the

Cold War. The region was in flux from the early 1990s onward, and, as a result, innovative approaches such as Agenda 21 had a better chance of coming to fruition there. Nevertheless, deficits remain, which concern not only the general goals, but also the multilevel approach taken by the agenda. On the one hand, nation states (and the European Union) clearly dominate the Baltic 21 network: important positions are assigned to representatives of national governments. On the other hand, Baltic 21 projects appear to be most successful when carried out by actors such as the UBC, which have particular competences and expertise in specific fields. The implementation of Agenda 21 is impeded by its strong reliance on national capacities, which are not evenly distributed among the countries in the region. The lack of national capacities may be compensated for by the capacities of transnational non-governmental actors, and Baltic 21 could become more successful through the even greater integration of actors such as the UBC.

Networks such as Baltic 21 and the UBC can develop capacities and instruments for implementation that cannot be created through intergovernmental cooperation alone. This fact should be emphasized with regard to local environmental policies in particular. By choosing the sustainable development of the Baltic Sea Region as an important goal for its organization, a general network organization such as the UBC can foster understanding of these issues among its various member cities. The UBC has developed a transnational identity of its own and has stronger ties with Brussels than with the national capitals in the region. By providing hands-on support and service for their member cities, a transnational network such as the UBC can complement the traditional modes of governance adopted by nation states – that is, international regimes and intergovernmental cooperation. However, nation states may be reluctant to support such transnational networks because the latter could affect and weaken the position of the former.

Our study clearly shows that new governance arrangements are influenced by the European Union, which has become a strong political player in the Baltic Sea Region, mainly through its direct involvement in international decision-making (e.g. HELCOM and Baltic 21), through European regulations (especially via directives) that aim at national governments, and through the funding of selected projects. Since sustainable development needs a broad base of support at all levels of government, Baltic 21's integrative approach seems to fit very well into the overall European strategy. Moreover, the EU has started to cooperate directly with transnational networks such as the UBC, and most of the funding for the UBC Environment Commission's projects is provided by the EU. The influence of the EU has been increasing steadily, and organizations such as the UBC have become players in the European multilevel system and have developed strong ties with Brussels.

In summary, it can be concluded that the achievement of sustainable development in the Baltic Sea Region undoubtedly requires a productive combination of national governance and new forms of governance beyond the nation state. In this respect, transnational policy networks, such as Baltic 21, and transnational networks, such as the UBC, represent promising new approaches that can complement the traditional cooperation between nation states via international regimes or intergovernmental cooperation. Moreover, governance of the Baltic Sea Region is becoming more and more embedded within European governance, which is leading to the Europeanization of the Baltic Sea Region.

NOTES

1 This was a result of Foreign Minister/Chancellor Willi Brandt's *Ostpolitik* (Eastern policy), which pushed for political and practical rapprochement towards the German Democratic Republic (GDR), and culminated in the signing of the *Grundlagenvertrag* (Treaty on the Basis of Intra-German Relations) in December 1972.

2 The signatories were Denmark, Finland, the Federal Republic of Germany, the German Democratic Republic, Poland, Sweden and the USSR.

3 Signatories to the revised Helsinki Convention of 1992 included all of the original contracting parties, with the reunited Germany legally succeeding the former FRG and GDR, plus the newly independent Baltic states: Estonia, Latvia and Lithuania.

4 The working groups are HELCOM MONAS (Monitoring and Assessment Group), HELCOM LAND (Land-based Pollution Group), HELCOM MARITIME (Maritime Group), HELCOM RESPONSE (Response Group), HELCOM HABITAT (Nature Protection and Biodiversity Group).

5 These statistics were provided by HELCOM itself (see www.helcom.fi/pollution/hazardous.html#achievements, accessed 19 September 2004).

6 The Convention for the Protection of the Marine Environment of the North-East Atlantic (OSPAR Convention) replaced the Oslo and Paris Conventions and entered into force in 1998.

7 Baltic 21 has taken some initial steps in this direction by contributing to the World Summit on Sustainable Development in Johannesburg, by playing a role in the European Environmental Ministers Conference in Kiev (Environment for Europe Process) in 2003, and by reaching out to initiate cooperation with the Euro-Mediterranean Region and the US (Baltic 21, 2003, p24).

8 The Europe-wide network 'Eurocities' was founded in 1986 on the initiative of Rotterdam, Barcelona, Frankfurt am Main, Milan and Lyon. Regional networks of cities and towns were created in the Mediterranean and Alpine regions in 1991 and 1996, respectively; on the 'Alliance in the Alps', see Amor (1999) and Behringer (2003).

9 Size and population are not exclusive criteria for entry into the UBC, as is the case with other municipal associations. In fact, its members include very small towns such as Bützow in Mecklenburg-Western Pommerania in Germany (population 9400) and Kärdla in Estonia (population 4100), as well as the largest cities in the Baltic Sea Region – namely, St Petersburg (population 4,730,000), Riga (population 790,000) and Stockholm (population 740,000). A glance at the statutes of the UBC shows that selection requirements are very liberal as 'any coastal city of the Baltic Sea and its gulfs, as well as any other city interested in the development of the Baltic Sea Region, may become a member city of the union.'

10   See UBC Statute, Articles 1, 2c and 2f (UBC, 2004, p93). Concerning the UBC strategy, sustainability is mentioned several times; see, for example, the section devoted to the 'UBC Agenda 21 Action Program' (UBC, 2004, p98).

11   In 1973, the heads of state and government assigned the then European Community certain competence in the field of environmental policy. With the 1986 Single European Act, European environmental policy was given a legal basis. Successively, with the treaties of Maastricht and Amsterdam, European competence was further extended. With the Treaty on European Union (TEU), qualified majority voting became the regular decision-making procedure in the Council of Environmental Ministers.

12   Communication from the Commission: Dialogue with Associations of Regional and Local Authorities on the Formulation of European Union Policy, COM (2003) 811 final, 19 December 2003.

13   With respect to Russia, nuclear safety is also a priority area.

REFERENCES

Amor, K. (1999) 'Das Gemeinde-Netzwerk "Allianz in der Alpen"' in: IFOK/ZKE (ed) *Was heißt hier Agenda? Analysen – Erfahrungen – Beispiele*, Dettelbach, Röll, pp77–185

Andersen, M. S. and Liefferink, D. (eds) (1997) *European Environmental Policy: The Pioneers*, Manchester University Press, Manchester and New York

Anderson, M. (1999) *Change and Continuity in Poland's Environmental Policy*, Springer, Berlin

Axelrod, R., Downie, D. and Vig, N. (eds) (2005) *The Global Environment: Institutions, Law, and Policy*, CQ Press, Washington, DC

Bache, I. and Flinders, M. (eds) (2005) *Multi-Level Governance*, Oxford University Press, Oxford

Baltic 21 (1998) 'Baltic 21 – An Agenda 21 for the Baltic Sea Region', *Baltic 21 Series*, no 1/1998

Baltic 21 (2000) 'Biennial Report – 2000', *Baltic 21 Series*, no 1/2000

Baltic 21 (2003) 'Baltic 21 Report 2000–2002: Towards sustainable development in the Baltic Sea Region', *Baltic 21 Series*, no 1/2003

Baltic 21 (2004) 'Five years of regional progress towards sustainable development: A Baltic 21 report to the prime ministers of the Baltic Sea states', *Baltic 21 Series*, no 1/2004

Behringer, J. (2003) 'Nationale und transnationale Netzwerke in der Alpenregion', WZB Discussion Paper no SPIV 2003-104, Wissenschaftszentrum Berlin für Sozialforschung, http://skylla.wz-berlin.de/pdf/2003/iv03-104.pdf

Benner, T., Reinicke, W. H. and Witte, J. M. (2003) 'Global public policy networks: Lessons learned and challenges ahead', *Brookings Review*, vol 21, pp18–21

Bruch, J. (1999) 'Umweltkooperation im Ostseeraum', PhD thesis, Johannes Gutenberg University of Mainz, Germany

Chasek, P., Downie, D. and Welsh Brown, J. (2006) *Global Environmental Politics*, Westview, Boulder, CO

Dorsch, P. (2003) 'Nationale und transnationale Vernetzung polnischer Städte und Regionen – Auf dem Weg zu einer nachhaltigen Stadt und Regionalentwicklung',

WZB Discussion Paper no SPIV 2003-106, Wissenschaftszentrum Berlin für Sozialforschung, http://skylla.wz-berlin.de/pdf/2003/iv03-106.pdf

Ehlers, P. (2001) 'Der Schutz der Ostsee – Ein Beitrag zur regionalen Zusammenarbeit', *Natur und Recht*, vol 23, pp661–666

Engström, A. (1998) 'How the union of the Baltic cities contributes to the development of the Baltic Sea Region', in C. Wellmann (ed) *From Town to Town: Local Authorities as Transnational Actors*, LIT-Verlag, Hamburg, pp171–183

European Commission (2001) *European Governance: A White Paper*, EC, Brussels

Fitzmaurice, M. (1992) *International Legal Problems of the Environmental Protection of the Baltic Sea*, Martinus Nijhoff/Graham and Trotman, Dordrecht, The Netherlands

Görmar, W. (1997) 'Erfahrungen und Perspektiven der transnationalen Kooperation zur Raumordnung in der Ostseeregion', *Informationen zur Raumentwicklung*, vol 6, pp405–418

Groth, N. B. (ed) (2001) *Cities and Networking: The Baltic Sea Region*, Report no 8-20012, Danish Centre for Forest, Landscape and Planning, Horsholm

HELCOM (2002) *Activities 2001*, Environment Proceedings no 84, Helsinki

Hubel, H. and Gänzle, S. (2002) 'Der Ostseerat. Neue Funktionen subregionaler Zusammenarbeit im Kontext der EU-Osterweiterung', *Aus Politik und Zeitgeschichte*, vol 19-20/2002, pp3–11

Jänicke, M. and Weldner, H., in collaboration with Jörgens, H. (eds) (1997) *National Environmental Policies: A Comparative Study of Capacity Building*, Springer, Berlin

Jann, W. (1993) 'Regieren im Netzwerk der Regionen – Das Beispiel Ostseeregion', in C. Böhret and G. Wewer (eds) *Regieren im 21: Jahrhundert zwischen Globalisierung und Regionalisierung*, Leske und Budrich, Opladen, pp187–206

Joas, M. (2003) 'Local Agenda 21 in the Baltic Sea Area: Ecological, economic and political stability for local level sustainable development', in L. Hedegaard, B. Lindström, P. Joenniemi, H. Eskelinen, K. Peschel and C.-E. Stalvant (eds) *The NEBI Yearbook 2003: North European and Baltic Sea Integration*, Springer and Nordregio, Berlin, pp111–126

Joas, M. and Hermanson, A.-S. (eds) (1999) *The Nordic Environments: Comparing Political, Administrative, and Policy Aspects*, Ashgate, Aldershot, UK

Jordan, A. (ed) (2005) *Environmental Policy in the European Union: Actors, Institutions and Processes*, 2nd edition, Earthscan, London

Jordan, A. and Lenschow, A. (2000) '"Greening" the European Union: What can be learned from the "leaders" of EU environmental policy?', *European Environment*, vol 10, pp109–120

Kern, K. (2001) 'Transnationale Städtenetzwerke in Europa', in E. Schröter (ed) *Empirische Policy- und Verwaltungsforschung. Lokale, nationale und internationale Perspektiven*, Leske und Budrich, Opladen, pp95–116

Kern, K. and Bulkeley, H. (2008) 'Cities, Europeanization and multi-level governance: Governing climate change through transnational municipal networks', *Journal of Common Market Studies*

Kindler, J. and Lintner, S. F. (1993) 'An action plan to clean up the Baltic', *Environment*, vol 35, no 8, pp7–31

Knill, C. (2003) *Europäische Umweltpolitik. Steuerungsprobleme und Regulierungsmuster in Mehrebenensystemen*, Leske und Budrich, Opladen

Lafferty, W. M. and Hovden, E. (2003) 'Environmental policy integration: Towards an analytical framework', *Environmental Politics*, vol 12, no 3, pp1–22

Lafferty, W. M. and Meadowcroft, J. (eds) (2000) *Implementing Sustainable Development: Strategies and Initiatives in High Consumption Societies*, Oxford University Press, Oxford

Lenschow, A. (2005) 'Environmental policy: "Contending dynamics of policy change"', in H. Wallace, W. Wallace and M. Pollack (ed.) *Policy-Making in the European Union*, 5th edition, Oxford University Press, Oxford, pp305–327

Lindström, Å. and Grönholm, B. (2002) *Progress and Trends in Local Agenda 21 Work within UBC Cities: Union of the Baltic Cities Local Agenda 21 Survey 2001*, Åbo Akademi University, Åbo

List, M. (1997) 'Das Regime zum Schutz der Ostsee', in T. Gehring and S. Oberthür (eds) *Internationale Umweltregime, Umweltschutz durch Verhandlungen und Regime*, Leske und Budrich, Opladen, pp133–146

Oberthür, S., Buck, M., Müller, S., Pfahl, S., Tarasofsky, R., Werksmann, J. and Palmer, A. (2002) *Participation of Non-Governmental Organisations in International Environmental Cooperation: Legal Basis and Practical Experience. UBA-Berichte 11/02*, Erich Schmidt Verlag, Berlin

Pierre, J. and Peters, B. G. (2000) *Governance, Politics and the State*, Macmillan/St Martin's Press, Basingstoke

Poutanen, E.-L. and Melvasalo, T. (1995) 'The Helsinki Commission and its ad hoc high level task force', in J. Köhn and U. Schiewer (eds) *The Future of the Baltic Sea: Ecology, Economics, Administration, and Teaching*, Metropolis-Verlag, Marburg

Reents, M., Krüger, C. and Libbe, J. (2002) *Dezentralisierung und Umweltverwaltungsstrukturen in Mittel- und Osteuropa. Ein Vergleich der EU-Beitrittsländer Polen, Tschechische Republik, Ungarn, Estland, Lettland und Litauen*, Deutsches Institut für Urbanistik, Berlin

Rosenau, J. (1999) 'Toward an ontology for global governance', in M. Hewson and T. J. Sinclair (eds) *Approaches to Global Governance Theory*, State University of New York Press, Albany, pp287–301

Rosenau, J. (2003) *Distant Proximities: Dynamics beyond Globalization*, Princeton University Press, Princeton

Stalvant, C.-E. (1999) 'The Council of Baltic Sea States', in A. Cottey (ed) *Subregional Cooperation in the New Europe. Building Security, Prosperity and Solidarity from the Barents to the Black Sea*, Macmillan/St Martin's Press, Basingstoke, pp46–68

Swedish Ministry of Environment (2000) *The Baltic – Our Common Sea*, Environment Report no 3, April, Swedish Ministry of Environment, Stockholm

UBC (Union of the Baltic Cities) (2001a) *UBC 10th Anniversary, 1991–2001: Past, Present, Future*, UBC Secretariat, Gdansk

UBC (2001b) *Report from the VI General Conference Social Justice in the Baltic Sea Region*, Rostock, Germany, 12–13 October 2001, UBC Secretariat, Gdansk

UBC (2002) *UBC Agenda 21 Action Programme*, UBC Secretariat, Gdansk

UBC (2004) *Report from the VII General Conference The Baltic Sea Wave – Business Development in the New Europe*, Klaipeda, Lithuania, 17–18 October 2003, UBC Secretariat, Gdansk

UN (United Nations) (1992) *Agenda 21: The United Nations Programme of Action from Rio*, UN, New York

VanDeveer, S. D. (1999) 'Capacity building efforts and international environmental cooperation in the Baltic and Mediterranean regions', in S. D. VanDeveer and G. D. Dabelko (eds) *Protecting Regional Seas: Developing Capacity and Fostering Cooperation in Europe*, Woodrow Wilson International Centre for Scholars, Washington, DC, pp8–32

Vartiainen, P. (1998) 'Urban networking as a learning process: An exploratory framework for transborder cooperation in the Baltic Sea Region', in U. Graute (ed) *Sustainable Development for Central and Eastern Europe: Spatial Development in the European Context*, Springer, Berlin, pp115–126

Varwick, J. (1998) 'Globalisierung und "Global Governance". Möglichkeiten und Missverständnisse bei der politischen Gestaltung des Globalisierungsprozesses', *Gegenwartskunde*, 47, pp47–59

Visions and Strategies around the Baltic Sea 2010 (VASAB 2010) (2001) Background Documents for VASAB 2010 PLUS, Spatial Development Action Programme, Essen

Voelzkow, H. (2000) 'Von der funktionalen Differenzierung zur Globalisierung: Neue Herausforderungen für die Demokratietheorie', in R. Werle and U. Schimank (eds) *Gesellschaftliche Komplexität und kollektive Handlungsfähigkeit*, Campus, Frankfurt am Main and New York, pp270–296

Wohlgemuth, K. (1998) 'Die "Union of the Baltic Cities" als Instrument der Förderung der Wirtschaftskooperation', in K.-H. Breitzmann (ed) *EU-Erweiterung im Ostseeraum und Kooperationsförderung durch Ostseeorganisationen – Herausforderungen und Chancen für Mecklenburg-Vorpommern*, Universität Rostock, Rostock, pp101–109

Young, O. (ed) (1999) *The Effectiveness of International Environmental Regimes: Causal Connections and Behavioral Mechanisms*, MIT Press, Cambridge, Massachusetts, and London

Internet sites

Helsinki Commission: www.helcom.fi/, accessed 4 October 2007
Baltic 21: www.baltic21.org/, accessed 4 October 2007
Union of the Baltic Cities: www.ubc.net/, accessed 4 October 2007

7

Local Governance for Sustainable Development: Local Agenda 21 in the Baltic Sea Region

Marko Joas

INTRODUCTION[1]

The paradigmatic shift from environmental policy-making through traditional government towards governance for sustainable development has, to some extent, changed the traditional patterns of political and administrative routines. This shift has changed the content of policies affecting the human–environment relationship to some (but still limited) extent. To a much higher extent, the shift has changed our view of how policy process should be conducted in the environmental policy sector. The shift implies a change from traditional top-down and representative government towards bottom-up and participatory governance.

One specific feature of the paradigmatic shift has included the introduction of several new policy tools within the environmental and sustainability toolbox. All new policy tools are considered soft in comparison with traditional legal and administrative regulatory tools. The new tools are often voluntary and based on common understanding of the problem and a wish to find common solutions. They include information-driven and economic instruments, as well as co-operative and participatory elements (see, for example Jordan et al, 2003).

Local Agenda 21 (LA21) is one of the 'new' instruments introduced after the Rio Earth Summit in 1992 (Joas and Grönholm, 2004, p500). It can be defined as a policy tool through which existing local structures are mobilized to meet sustainability challenges from local and global environments, as well as local social and economic conditions, and, to some extent, to create new structures among political and non-political actors. These new structures are based on networks and thus meet the definitions of multilevel governance; even if they are based on existing structures, they also imply changes.

Earlier research indicates that the Baltic Sea Region is an interesting arena to study the shift at the local level towards sustainability governance. First of all, the Baltic Sea is a vulnerable natural environment, with a limited range of natural biodiversity; a developing aquatic ecosystem (Bonsdorff, 2006, p383); and suffers from environmental pressures stemming from the volume of resource use in agriculture and vehicle use in transport.

Second, the deep political cleavage following the former Cold War runs through the Baltic Sea between East European and West European countries (see, for example, Archer, 2003, for the security policy change in the region). This has resulted in generalized differences in political and economic capacities and resources, as well as particular gaps with regard to environmental policies. This social and economic cleavage is still, to some extent, visible in political and administrative structures in transition countries (see, for example, King, 2003, p55) and, consequently, also affects the possibility of conducting sustainability policies. As described, for example, in Joas (2003), however, new political structures for implementing environmental and sustainability tools and policies are visible. This applies within countries, but also between nations: new international, but still sub-national, cooperation structures have entered the policy-making arena. This development is evident in the Baltic Sea Region (see, for example, Kern and Löffelsend, 2004).

From a European perspective, the introduction of LA21 to the Baltic Sea Region has been widely seen as a success story, with the new tool diffusing rapidly from Nordic countries to the Baltic republics and Poland with the help of European Union (EU) support (Cameron and Joas, 2007), as well as to Russia and other non-EU members. This chapter discusses whether this development is genuine or just a question of good marketing by comparing Baltic transition economy countries with Nordic countries, and these groups further with other transition economy Central and Eastern European (CEE) countries and other West European countries.

This chapter assumes that the emerging new multilevel governance structures in the Baltic area (i.e. extensive use of networking in the region) can at least partly be seen as an explanation for any activity difference. To test this assumption, two sets of comparative data are used. First, data was collected on overall LA21 implementation activity at the national level from UN and other similar existing databases. This basic data was, to a large extent, collected in 2003 from UN Rio+10 meeting national reports; national descriptions were updated during 2005 and 2007. In addition, a comparative database from 138 European cities active in an LA21 or similar sustainability policy process was also used. This Local Authorities Self-Assessment of Local Agenda 21 – project (LASALA) database is collected from self-evaluations by cities all over Europe within a project finalized in 2001.[2]

FROM GOVERNMENT TO GOVERNANCE

The introduction of governance as a catchword for social research, as well as policy documents, is, to a great extent, connected with the shift in the view of what governments are expected to do. The seminal volume by Osborne and Gaebler (1993) called for governments to change their behaviour from producing to controlling services. This development had, of course, already been empirically visible for a longer period of time as a consequence of changes in political attitudes and economic development. For governments, this change also meant that a growing number of actors were to be involved in the business of government (e.g. private, competing and non-governmental service producers). 'Governing issues generally are not just public or private, they are frequently shared' (Kooiman, 2003, p3) is one vision for this development.

Another development trend pointing towards a change in our view of government has been the value base of post-material citizens, from prioritizing economic welfare to prioritizing social, intellectual and aesthetic goals (Papadakis, 1998, p146). Individualism is often seen as one of the key factors changing the relationship between citizens and government towards solutions that take different individuals and their needs into account instead of mainstreaming all input to, and output from, government.

As a concept, governance is used to describe empirical efforts by governments in order to adapt themselves to a changing society, but also as a theoretical concept highlighting the changing role of the state in the coordination process of different social systems (Pierre, 2000b, p3). Therefore, the definitions and usage of governance have changed significantly over time, following empirical evidence of frequently changing practices. The key issue at stake is, for most scholars, the changing relationship between different stakeholders (and among them the public) and more stable government structures. Björk et al (2003, pp112–113) highlight three separate views on governance: new forms of institutionalized structures, new forms of governing processes and the results of these processes. They continue to count up to no less than seven different forms of governance, all studied by contemporary scholars: the minimal state; new public management; good governance; the socio-cybernetic system; self-organizing networks; multilevel governance; and public governance (Björk et al, 2003, p113). All of these are used frequently in political and social analysis, thus highlighting governance as a change in the processes of interaction between different political actors, and this despite the political and administrative level in focus. In addition, governance is also used as a concept in business administration (corporate governance) and economics.

One line of argument has been to interpret governance as an 'erosion of traditional bases of political power' (Pierre, 2000b, p1). This is not, however, the main conclusion for many other authors. The erosion of traditional government in favour of governance can, in fact, be seen just as an additional tool for governments to achieve their political goals that would be impossible to realize (with reasonable costs) with traditional tools. This means that governance can be 'explained by an awareness that governments are not the only actors addressing major societal issues; that besides the traditional ones, new modes of governance are needed to tackle these issues; that governing arrangements will differ from global to local and will vary from sector to sector' (Kooiman, 2003, p3). National governments are not disappearing from the stage; their role is just changing from acting to steering and cooperating with other actors (Eckerberg and Joas, 2004, p406). This changing role towards cooperation can, if we look at the local government and local public administration sphere within the environmental policy sector in a cross-sectoral sustainability setting, be seen as a major success factor for sustainability policies based on cooperation between government and civil society (Evans et al, 2005).

This development has appeared at all sociatal levels, but is most obvious at the local level: 'The change from traditional local government to a more complex network of agencies involved in "local governance" is no longer theory. It has become practice' (Goss, 2001, p1). Due to the complexity of the actor networks within the sustainability policy sector, local authorities are gaining in power, broadening the freedom of movement both within the country and outside national borders. On the one hand, local authorities' autonomous position is constrained by several different local-level actors, who are often different to the actors with access to the 'normal' local policy-making process. The scope for the political process at the local level, therefore, is becoming wider and more open. On the other hand, other political units beyond national governments can influence environmental and other policy processes at the local level, through sub-governmental transnational networks and international organizations (Joas et al, 2007). In fact, international contacts within the specific policy field of sustainable development seem to be of particular interest for local authorities (Joas, 2001, p261).

Analysing this multilevel governance phenomenon in the Baltic Sea Region is demanding but, at the same time, highly rewarding on several grounds. First, the traditional Nordic welfare model that places local government in an autonomous but responsible position in its interaction with the public can be defined as a manifestation of governance (Baldersheim and Ståhlberg, 2002). Indeed, there is a long tradition of governance in the area. Second, the 'policy-making processes' in the Baltic Sea Region seem to be well suited to analyses of multilevel governance since several policy networks with different actors are active in the region. In fact, even if the bulk of these actors are new, many of them are also deeply rooted in the region. Local authorities are also active within various networks acting in the region's policy field (Kern and Löffelsend, 2004; see also Joas and Grönholm, 1999).

LA21 AS A POLICY TOOL SUITABLE FOR NETWORKS

The local authority level of government has been actively promoting sustainable development throughout the Baltic region ever since the United Nations Conference on Environment and Development (UNCED) in Rio de Janeiro in 1992. Local Agenda 21 evolved from Agenda 21, a non-binding plan for sustainable development signed by more than 170 countries at the summit. Agenda 21 contained 40 chapters. The shortest chapter from Agenda 21, Chapter 28, encouraged local authorities throughout the world to conduct their own Local Agenda 21 action plans for a more sustainable future – both from a local and global perspective. The tenth anniversary of Rio, the World Summit on Sustainable Development in Johannesburg in 2002, identified local government as one of the key agents in the overall development towards sustainability, even though lack of action is still evident.

Local Agenda 21 is primarily defined as a voluntary policy tool with the main goal of activating already existing structures, but also, to some extent, of creating new institutions, which has also happened in a surprisingly high number of cities (Joas and Gronholm, 2004, p 500, see also Joas, 2003). This activity is planned to take the local communities towards a higher level of sustainability (i.e. to meet the challenges from local environmental, social and economic conditions).

Earlier empirical studies have highlighted the importance of networking activities, with local-level development clearly spurred on by cooperation within local communities and outside local governmental and national borders (see, for example, Joas, 2000, for an empirical study of LA21 diffusion within the region).

THE EMPIRICAL TASK

The way in which LA21 is used as a tool varies to a significant degree between different political systems. This is, of course, partly dependent upon differences in the position of local authorities in varying countries. The basic resources needed for local government to act also vary according to general economic conditions. Even within one nation, a significant variation can be seen in the content and scope of the local-level Agenda 21 process. Thus, different structures for implementing LA21 are visible not only between nations, but also within countries. At least in the Baltic Sea Region, an additional input to the diffusion of LA21 is given by the activity of numerous regional transnational cooperation structures between sub-national actors. This is tested as one possible independent variable to explain differences in LA21 activity levels in the Baltic compared to the rest of Europe.

Other explaining variables are not the main subject of interest in this chapter; at least one other additional variable is also discussed: the existence of national coordination bodies for sustainable development. This was seen as a key factor in explaining LA21 implementation success in the study by Lafferty and Eckerberg (1998, p253).

LA21 activity is measured in two different ways. First, this chapter summarizes the overall level of LA21 activities in the different groups of European countries (i.e. the number of LA21 processes and their existing national-level support mechanisms). Second, the scope and content of a selection of existing LA21 processes are analysed as one indicator. The main regional interest is for all examples in the Nordic and Baltic regional setting; but all data are presented according to a broad comparative perspective, including figures for other parts of Europe as well.

Data presentation

Data in this chapter are based on two parallel data collection processes. First, a basic knowledge of Local Agenda 21 has been collected from several different sources. Earlier studies of national-level LA21 activities within the region have been cited, and information was also collected through several key-person interviews in the Baltic countries and Russia. In addition, national follow-up reports to the Rio+10 process were analysed through the UN database (see www.un.org/esa/agenda21/, also reported in Joas, 2003).

Second, data from the EU-funded research project LASALA, which included a large survey of cities all over Europe, were analysed (see Evans and Theobald, 2001, as well as Grönholm et al, 2001). Local authorities involved in LASALA underwent an internet-based self-evaluation process. A major part of this study included a comprehensive survey that was answered by the local government LA21 coordinator. In all, LASALA attracted 228 participants from 29 European countries, while 146 local authorities completed the survey. The following analysis draws upon the answers from 138 local authorities (i.e. answers that are of such quality that a reference value for different process qualities was measurable). The regional division of responses to the LASALA survey was deemed to be satisfactory. The Baltic Sea Region was very active in participating in the LASALA process, even though a majority of these answers still came from the Nordic countries and Germany.

Local Agenda 21 activities

Local Agenda 21 began as an activity by and for the rich industrialized and developed part of the world. When participating in the United Nations Rio+5 meeting in New York in 1997, the International Council for Local Environmental Initiatives (ICLEI) demonstrated that local authorities in 64 countries were involved in LA21 activities, and the number of active local authorities had reached

1800 (Joas, 2003). A second worldwide survey by ICLEI and the UN for the Rio+10 meeting, directed at both national organizations for local government as well as all interested local authorities, confirmed the success of LA21 as a policy tool for local government. During 2002, LA21 processes were conducted in 113 countries, a number that had doubled in only five years. All in all, 6416 processes were reported to the UN survey, a number more than three times higher than in 1997. A total of 5292 of these processes (i.e. 82 per cent) were in Europe. National campaigns promoting LA21 are also present in Europe, where the emphasis on activity is still in the Nordic and Western European countries. Germany leads the way, both in the number of participating local authorities and in the number of inhabitants covered by an LA21 process. LA21 is also widespread throughout the Nordic countries. The second UN survey found 2042 LA21 processes in Germany and 1128 processes in the Nordic countries. This is about half of all LA21 processes in the world (*Second Local Agenda 21 Survey*, 2002).

Traditionally, the Nordic countries and Germany have been seen as forerunners in environmental policies (see, for example, Andersen and Liefferink, 1997); therefore, it is only natural to find the Nordic countries leading in LA21 activities. A similar logic also applies for Germany. These assumptions are verified at least to some extent by earlier research (see Lafferty and Eckerberg, 1998; Evans and Theobald, 2001; Grönholm et al, 2001).

LA21 IN THE BALTIC TRANSITION ECONOMY COUNTRIES[3]

In this analysis we include the Baltic group of countries, beginning with the Baltic republics of Estonia, Latvia and Lithuania. As seen within a political, geographical and economical context, Poland and the Russian Federation will also be discussed as countries belonging to this group. All these countries, except Russia, are today members of the EU, a fact indicating rapid economic, political and administrative progress.

The overall level of LA21 activity is clearly lower for this group of countries, compared with Nordic and Western Europe countries. The total number of local authorities implementing LA21 across the whole region is approximately 147, and half of these processes are to be found in Poland. Activities are clearly lower both in the Baltic countries and in Russia. The number of Local Agenda 21 activities was, however, still increasing at the time of data collection, a developmental trend very different from the rest of Europe (see, for example, Joas, 2003, p115). It must be emphasized that there are also clear forerunners among the Baltic cities that easily place themselves among the forerunners in our group of Nordic and German local authorities. This will be analysed in the second stage of the empirical analysis.

It is also obvious that the implementation rate for LA21 is increasing rapidly in the group of Baltic countries. A 1998 survey directed at the member cities of the Union of the Baltic Cities (UBC) showed that only a handful of Baltic member

Table 7.1 *Level of LA21 activity in the Baltic Sea Region*

Region	Country	National coordination on sustainable development	National campaign for LA21	Number of LA21 processes	Number of LASALA cases	Percentage of LASALA cases
Baltic	Estonia	Yes	No	29	3	2.2
Region	Latvia	EQ yes	No	5	3	2.2
Countries	Lithuania	Yes	No	14	2	1.4
	Poland	Yes	No	70	2	1.4
	Russia	EQ yes	No	29	3	2.2
	Baltic total	5/5 = 100%	0/5 = 0%	147	13	9.4

Note: EQ = equal body for coordination.
Source: Second LA21 Survey: National Implementation of Agenda 21: A Report, and LASALA database (n = 138)

cities in the UBC network had commenced LA21 work in 1998, but almost half of the CEE local authorities in that study planned to do so (see Grönholm and Joas, 1999). During 2001, the number of active LA21 processes in the Baltic group was about five times higher (see Lindström and Grönholm, 2002).

Estonia began working with sustainability policies soon after regaining independence in 1991. The 1995 Act on Sustainable Development in Estonia to some extent followed Agenda 21 topics. In addition, a national coordination body for sustainable development was created in Estonia, strengthening cross-sectoral implementation between the administrative bodies within the national administration. This politically high-level Commission on Sustainable Development includes 27 members representing the government, various ministries and the scientific community. Estonia has also lately redefined goals for sustainable development: in 2005, the new Estonian National Strategy on Sustainable Development (Sustainable Estonia 21) was approved by the Estonian parliament.

The Estonian government also had plans to finance local and regional Agenda 21 activities; but most of these activities are still, in practice, voluntary, often supported only by transnational networks, as well as by twinning arrangements. No national LA21 campaign exists, according to the UN survey. Nevertheless, at least 29 Local Agenda 21 or similar processes were identified in the 2002 survey (*Second Local Agenda 21 Survey*, 2002; *National Implementation of Agenda 21: A Report*, 2002; Joas, 2003).

In 2003, the Association of Estonian Cities estimated that existing LA21 processes are initiated, as well as funded, to a large extent by local authorities themselves. Knowledge of this process comes from several sources: international organizations and partners, local government organizations and, naturally, from within cities themselves. Of the total 33 Estonian cities and towns and 194 municipalities, only three had, by 2007, finalized long-term LA21 action plans

(Kuressaare, 1997; Tartu, 1998; and Viljandi, 2002), and an additional four larger cities (Pärnu, Jõgeva, Valga and Narva) were active within an LA21 process. Several other local authorities, including the capital Tallinn, actively participate in a similar process that can be defined as an LA21 process. Since most large towns are working with sustainability issues, a majority of the population can be said to live in municipalities with an ongoing LA21 process. However, only four local authorities had, by spring 2007, signed the Aalborg commitments (Joas, 2003; Alakivi, 2007).

In Latvia, the overall LA21 situation is comparable to the situation in Estonia. During 2003, approximately 20 cities and local authorities of the total 7 major cities, 56 towns and 473 other forms of local authorities were involved in an LA21 process; however, in the UN survey only five active processes were identified. Data have not been updated since then. Two of the Latvian LA21 processes, in Riga and Jûrmala, have managed to finalize an LA21 action plan, and a further three towns are working on an individual LA21 process (Rêzekne, Rûjiena and Talsi). In addition, there is a larger regional LA21 process in Bärtava that includes eight small rural municipalities. Otherwise named sustainability processes, with similar goals but different means, are available in a further six cities. Close to half of the Latvian population lives in cities that are already included in some form of process aimed at sustainable development (Joas, 2003).

According to the Latvian government, reporting to the UN, there is some level of national backup to the Latvian Agenda 21 process that takes the form of a national coordinating body for the sustainability process. Regional environmental boards with responsibility for sustainable development policies give some governmental backup to local authorities. Nevertheless, in Latvia, local-level input for LA21 processes seems to be of central importance (*National Implementation of Agenda 21*, 2002; Joas, 2003).

Lithuania is divided into 60 local authorities; of these, 16 large local authorities are involved in an LA21 process. This means that more than 40 per cent of the total population lives in municipalities that are involved in an LA21 process. The most active agents are found at the local level. The interviewee estimated that as many as three out of every four LA21 processes are initiated by local actors even if the role played by international non-governmental organizations (NGOs) is also evident. Four ongoing projects are recognized as forerunners in Lithuania and, to some extent, internationally: LA21 processes in Kaunas, Klaipeda, Panevezys and Visaginas. In addition, 12 smaller local authorities were active in an NGO (ECAT-Lithuania) led cooperation process (Joas, 2003).

According to our interview, national government backup for local-level activities seems to be minimal. No national coordination was provided for LA21, according to the *Second Local Agenda 21 Survey* (2002); but there is a national-level Agenda 21 coordination body, according to the Lithuanian government's report to the UN, updated in 2007. The role of local authority activity is more proactive (*National Implementation of Agenda 21*, 2002; Joas, 2003).

Information from the Russian Federation is, to some extent, confusing. When interviewed in 2001, the Union of Russian Cities estimated that from the rough total of 2500 local authorities within the federation, more than 10 per cent are involved in a sustainability process. Between 100 and 200 of these processes could be directly defined as LA21. But, according to the *Second Local Agenda 21 Survey* (2002), the number of LA21 processes in Russia was confirmed to be only 29. The Russian Federation reporting to the UN regarding sustainable development is also very modest in extent.

LA21 activity level seems to be somewhat higher in the Baltic Sea Region, largely due to a higher level of awareness created through active networking in the region. Local and national-level political pressure from other countries in the Baltic region can also play a role in this respect. A majority of the processes are activated, funded and conducted by local authorities; the cities thus act on their own. Many of the larger Russian cities in the region – for example, St Petersburg, Novgorod and Kaliningrad – are active. Much of the progress also depends upon foreign support, as national funding and support are almost non-existent despite the existence of a national coordination body for sustainable development, providing limited national funding (*National Implementation of Agenda 21*, 2002; Joas, 2003).

Poland seems to be the most advanced Baltic transition economy country in an LA21 comparison. Even when compared to the rest of Europe, some Polish cities (e.g. Gdansk) are regarded as forerunners in local sustainability policies. Already in 1994, the first national coordinating Committee for Sustainable Development was created, and it was further developed in 1997 to include regional sustainable development. No specific national campaign for LA21 exists, however. Poland stresses in its Rio process reporting that local authorities are self-governing regarding sustainability policies. Polish local authorities, for example, prepare local physical development plans of their own. Johannesburg data reported that at least 70 LA21 processes are active in Poland. This is a remarkable advancement from 1997, when only 11 LA21 processes were reported to the UN. Nevertheless, this still represents a minority when viewed against the 265 cities and towns in Poland, and the more than 2000 rural local communities in the country. The responsibility for further action is still in the hands of local government, often supported by international and national NGOs (*Second Local Agenda 21 Survey*, 2002; Joas, 2003).

LA21 ACTIVITY IN THE NORDIC COUNTRIES[4]

The Nordic countries – Finland, Sweden, Denmark, Norway and Iceland – are all, as indicated earlier, forerunners in environmental policies. Even if Norway's and Iceland's natural influence is limited in a Baltic context, their cultural ties are so important that these countries must be briefly analysed.

Eckerberg et al (1999) considered Sweden to be the Scandinavian and world-wide pioneer in initiating LA21 activities.

Table 7.2 *Level of LA21 activity in the Nordic countries*

Region	Country	National coordination on sustainable development	National campaign for LA21	Number of LA21 processes	Number of LASALA cases	Percentage of LASALA cases
Nordic	Denmark	EQ yes	Yes	216	1	0.7
Countries	Finland	Yes	Yes	303	8	5.8
	Iceland	EQ yes	Yes	37	1	0.7
	Norway	EQ yes	Yes	283	6	4.3
	Sweden	Yes	Yes	289	11	8.0
	Nordic total	5/5 = 100%	5/5 = 100%	1128	27	19.6

Note: EQ = equal body for coordination.
Source: Second LA21 Survey: National Implementation of Agenda 21: A Report, and LASALA database (n = 138)

From an early stage, Sweden had a national administrative body to coordinate Local Agenda 21 activities. Thus, already by 1997, all 288 local authorities in Sweden had initiated an LA21 process. By 2004 more than 70 per cent of all Swedish local authorities could show a politically accepted action plan (Dahlgren and Eckerberg, 2005). Central economic investments (e.g. through the Swedish Local Investment Programme for sustainable development (LIP) and Climate Investment Programme (KLIMP) funding schemes) from the national government were the main reason for this high level of activity (Holm et al, 2005). Since the late 1990s, however, there has been a visible decrease in LA21 activity in Sweden. In many local authorities, LA21 activities were perceived as a project, rather than an ongoing process. Resources for local-level coordination have, for example, been cut substantially. In 2001, 65 per cent of all municipalities employed staff for LA21 coordination and 69 per cent had devoted funding for these activities. By 2004, both these figures were around 50 per cent. In general, however, Sweden still must be considered a pioneer in local-level sustainable development even if the forms and headings of local activities might be changing (Holm et al, 2005).

In the rest of the Nordic countries, the level of LA21 activity has been lower since interest in the process is more recent. Since 1987, Finland has had a national Commission for Sustainable Development, with a separate subdivision for local government. The Association of Finnish Local and Regional Authorities (AFLRA) has often been the main driving force for local-level action, in cooperation with national ministries. All LA21 processes in Finland have been voluntary, and the strongest support has been in the form of know-how delivered by AFLRA, often cooperating with the Ministry of the Environment. In 2005, a total of 310 out of 444 Finnish municipalities stated that they have initiated a Local Agenda 21 process; in practice, not more than 50 per cent of all municipalities can be said to have been actively working with LA21 as a policy tool for local-level sustainability (confirmed by an Åbo Akademi Survey, 2004; see Holm et al, 2005). Since all larger cities have been active, close to 83 per cent of the national population is living in a municipality with an ongoing or finalized LA21 process (Holm et al, 2005; see also www.kunnat.net).

In Norway, comprehensive environmental administrations were introduced to all Norwegian municipalities in the early 1990s (known as the MIK reform). This was considered to be a sufficient LA21 contribution. National coordinating activities were also limited until the late 1990s, when an LA21 unit was established within the Ministry of the Environment. Since then, however, Norwegian local authorities have initiated LA21 processes. In 1997, there were approximately 60 to 70 LA21 processes in Norway; but this has rapidly increased. During the period of 2000 to 2002, close to 300 Norwegian municipalities were involved in LA21 activities, representing about 69 per cent of all Norwegian municipalities. Norwegian local authorities have committed themselves to sustainability by signing a national declaration for sustainability – the Fredrikstad Declaration (in early 2002, a total of 263 out of 435 local governments had signed it). Since 2002, almost all supporting measures for local LA21 processes, however, have been terminated. This has meant that the number of active LA21 processes has started to rapidly fall; and by 2004, close to one third of all municipalities stated that their process was terminated, leaving only 38 per cent of municipalities that continue these activities (Joas, 2003; Holm et al, 2005).

During the mid 1990s, the level of LA21 activity was also lower in Denmark than in Sweden. However, the overall activity level can still be seen as rather high: by 1998 up to 68 per cent and in 2001 up to 75 per cent of Danish municipalities responded that they had LA21 activities in progress. At the same time, an additional 13 out of 14 regional authorities worked with LA21. Today, with the passing of a law requiring municipalities to proceed with mandatory LA21 planning by the Danish parliament, it is clear that all municipalities will adopt the process (Joas, 2003; Holm et al, 2005). The number of local government units was also drastically reduced in Denmark in 2007, with only 98 local authorities.

In Iceland, the national Agenda 21 process began in 1998. The main focus was on Local Agenda 21. Development has been rapid, and by 2001 up to 50 out of a total of 124 local authorities in Iceland were already involved in a nationally coordinated LA21 process. Although this is only about 40 per cent of all municipalities, it nonetheless represents more than 90 per cent of the population as the Icelandic population is concentrated in the capital area (Joas, 2003; see also www.samband.is/dagskra21).

LEVEL OF LA21 ACTIVITY IN OTHER EUROPEAN COUNTRIES

To be able to estimate the overall Local Agenda 21 activity level in European transition economy countries as the natural comparison group for the Baltic transition economy countries, this study has had to depend upon self-reporting from these countries to the UN Rio+10 process. Only general findings are summarized here.

In all, 175 Local Agenda 21 processes are to be found, according to the *Second Local Agenda 21 Survey* (2002), in the group of other CEE countries. This is only slightly higher than the number of recognized LA21 processes in the Baltic region transition economy countries. In terms of the level of LA21 implementation, the group of other CEE countries is, however, very heterogeneous. At least two groups of countries can be traced. First, there are a number of countries that are still in the process of building enduring structures for local government and thus are not yet able to go deeper into the content of different policy sectors, such as sustainable development. Second, there is a group of countries that are evidently already taking the first steps towards a modernized local government structure and that are also able to begin sustainability processes of their own. These countries are often either new accession countries and are therefore in the same situation as most of the Baltic region EU members or countries.

Countries that are clearly lagging behind are Armenia, Azerbaijan, Belarus, Bosnia, Georgia and FYR Macedonia. All of these countries are still healing wounds after internal or external conflicts, and are still building the structures of local governance. A few of these lack even a national coordination body for sustainable development. Several countries are close to taking a step towards the next level of implementation; these include Albania, with at least seven reported LA21 processes and some level of governmental support; Moldova, which reported in its country report approximately five ongoing processes within a national campaign, although this evidently has not been reported in UN conclusion reports (see unavailable data in Table 7.3); Hungary, which according to the level of development should be able to show more than just nine active processes (other similar processes are present, as well as government support); Slovenia, where national coordination for LA21 was just recently introduced after a reorganization of the entire local government system; and, finally, Ukraine, which has submitted no information to the UN on this subject, but which can report a handful of active LA21 processes. EU membership or expected EU membership will, however, change the pace of reform.

The second group of countries clearly performs much better. Almost all ongoing LA21 processes are found in just five to six countries in this group. According to the *Second Local Agenda 21 Survey* (2002), most active local authorities are found in the Czech Republic (42) and the Slovak Republic (30). The Czech Republic claims to have 57 local and regional LA21 (or similar) processes since 1998, and further claims national legislation promoting LA21 as a policy tool. The support for local action is clearly lower in the Slovak Republic; but there are also steps taken towards coordination of regional LA21 efforts.

In addition, several active LA21 processes are reported by Romania, Yugoslavia (today Serbia), Bulgaria and Croatia. All of these report several LA21 processes

Table 7.3 *Level of LA21 activity in the European transition economy countries*

Region	Country	National coordination on sustainable development	National campaign for LA21	Number of LA21 processes	Number of LASALA cases	Percentage of LASALA cases
Other transition economy countries	Albania	EQ yes	No	7	n.a.	n.a.
	Armenia	Yes	n.a.	n.a.	n.a.	n.a.
	Azerbaijan	No	n.a.	n.a.	n.a.	n.a.
	Belarus	Yes	n.a.	n.a.	n.a.	n.a.
	Bosnia	No	No	1	n.a.	n.a.
	Bulgaria	Yes	No	22	2	1.4
	Croatia	EQ yes	No	20	2	1.4
	Czech Republic	Yes	No	42	4	2.9
	Georgia	Yes	n.a.	n.a.	n.a.	n.a.
	Hungary	Yes	No	9	n.a.	n.a.
	FYR Macedonia	EQ yes	n.a.	n.a.	n.a.	n.a.
	Moldova	EQ yes	n.a.	n.a.	n.a.	n.a.
	Romania	Yes	No	12	7	5.1
	Slovak Republic	Yes	No	30	1	0.7
	Slovenia	Yes	No	3	n.a.	n.a.
	Ukraine	EQ yes	No	9	1	0.7
	Yugoslavia	In progress	No	20	n.a.	n.a.
	CEE total	14/17 = 82%	0/11 = 0%	175	17	12.3

Notes: EQ = equal body for coordination; n.a. = not available.
Source: Second LA21 Survey: National Implementation of Agenda 21: A Report, and LASALA database (n = 138)

(see Table 7.3). Yugoslavia has reported that after a long period of an unstable political situation, the build-up towards local governance is just beginning. The implementation of LA21, or Local Environmental Action Planning, is mostly supported by an NGO (Regional Environmental Centre). Bulgaria has reported some level of governmental support for LA21, perhaps resulting from the fact that 22 processes are currently active. Croatia has not reported on the issue of LA21 to the UN; but in the survey, at least 20 LA21 processes can be found, two of which participated in the LASALA self-evaluation project. Finally, Romanian LA21 processes are coordinated within the national sustainable development steering committee and supported by international efforts. Nine out of the 12 reported LA21 projects are also participating in some form of a common pilot project for LA21. Romanian interest in participating in the LASALA project was also evident.

Finally, without going into detail, it can be stated that Western Europe, especially Germany, The Netherlands and (earlier) the UK, can be classified as forerunners in introducing sustainability policies to the local government level.

Table 7.4 *Level of LA21 activity in Southern and Western European countries*

Region	Country	National coordination on sustainable development	National campaign for LA21	Number of LA21 processes	Number of LASALA cases	Percentage of LASALA cases
Western Europe	Austria	Yes	No	64	2	1.4
	Belgium	Yes	No	106	n.a.	n.a.
	Germany	EQ yes	No	2042	17	12.3
	Ireland	Yes	Yes	29	1	0.7
	Luxembourg	In progress	No	69	n.a.	n.a.
	The Netherlands	EQ yes	No	100	1	0.7
	Switzerland	EQ yes	No	83	n.a.	n.a.
	UK (including Northern Ireland)	EQ yes	Yes	425	26	18.8
	Western Europe total	7/8 = 88%	2/7 = 29%	2918	47	34.0
Southern Europe	Cyprus	n.a.	No	1	n.a.	n.a.
	France	Yes	No	69	3	2.2
	Greece	In progress	No	39	1	0.7
	Italy	Yes	Yes	429	20	14.5
	Malta	Yes	n.a.	n.a.	n.a.	n.a.
	Portugal	EQ yes	No	27	3	2.2
	Spain	EQ yes	No	359	7	5.1
	Southern Europe total	5/7 = 71%	1/6 = 17%	924	34	24.6

Notes: EQ = equal body for coordination.
The following small nations are not listed in this table: Andorra; Monaco; San Marino (according to the UN, this nation is not implementing LA21); and Liechtenstein, which has reported no data to the survey, but is, according to the UN, implementing LA21.
Source: Second LA21 Survey: National Implementation of Agenda 21: A Report, and LASALA database (n = 138)

However, the UK seems to be experiencing set backs, while activity levels in Italy and in Spain seem to be high. The highest number of new signatories to the *Aalborg Charter*, for example, comes from Italy, based on an option to obtain additional national funding for local sustainable development action.

COMPARING NATIONAL AND LOCAL LA21
IMPLEMENTATION ACTIVITIES

To conclude the comparison between the levels of LA21 implementation in the two research groups, data from Tables 7.1 to 7.4 are collated in Table 7.5.

Table 7.5 *An analysis of the breadth of LA21 activity*

Region	Number of reported LA21 processes	National coordination on sustainable development	National coordination on sustainable development (percentage)	National campaign for LA21	National campaign for LA21 (percentage)
Nordic	1128	5/5	100	5/5	100
Baltic	147	5/5	100	0/5	0
Other transition economy countries	175	14/17	82	0/11	0
Western Europe	2918	7/8	88	2/7	29
Southern Europe	924	5/7	71	1/6	17
Total	5292	36/42	86	8/34	24

Source: Second LA21 Survey: National Implementation of Agenda 21: A Report, and LASALA database (n = 138)

When counted by numbers of active and reported LA21 processes, it is clear that the two regions are very similar. However, the activities seem to be more evenly divided within all nations in the Baltic transition economy group. In other CEE countries, several nations hardly show any interest in LA21 or are unable to report any ongoing LA21 processes. For both groups, the UN reports that no LA21 campaigns exist on a national level. All Baltic transition economy countries have a national coordination body; only three countries in the group of other CEE countries lack such a body.

In sum, measured as overall LA21 implementation activity, no significant differences between the two groups of countries were found.

Scope and content of LA21 processes: An empirical analysis of LASALA data in the research region

The second indicator of LA21 implementation success is the scope and content of the reported LA21 processes. Seen from a process perspective, how well have the local Agenda 21 processes developed in different countries? What is the quality of the projects, not just the measured quantity, within the research regions?

In order to measure this, an earlier tested indicator of process qualities for LA21 processes was used. In the EU-funded LASALA project, which ran from 2000 to 2001, a tool was created to evaluate LA21 processes in a holistic and process-oriented way (known as 'the LASALA good practice identification model'). This model is based on the Public Administration Excellence Model, and was developed further in order to meet the needs and challenges addressed by the intentions and the specific character of the Agenda 21 document (see details in Grönholm et al, 2001).

The nine criteria used in the identification process were divided into factors that can be seen as enablers for a successful LA21 process and factors that expected results from an LA21 process.[5] The scoring for each of these nine criteria is based on indexes created by the research consortium, from a 20-page self-evaluation survey returned by a participating local government LA21 coordinator. The participating cities and local governments were self-selected since the project was a self-evaluation exercise.

The different criteria were weighted at the end to form a total index, with a maximum of 100 points possible. Differences among participants could easily be found even if there was an evident concentration just below a 50 per cent level of goal achievement.

Table 7.6 *The scope and content of LA21 processes based on LASALA good practice self-evaluation mean scores for different regions*

Region	Mean LASALA score	Sample population size	Standard deviation
Western Europe	45.65	50	9.597
Southern Europe	39.42	31	15.437
Nordic Countries	46.10	27	13.118
Baltic Countries	46.36	13	11.590
Other CEE Countries	40.90	17	11.023
Total	43.82	138	12.330

Source: LASALA database

Table 7.6 presents the mean values from the LASALA good practice scoring following the regional comparison discussed in this chapter. The assessed processes, on average, show little difference in Local Agenda 21 activity levels between Nordic and Baltic countries. The mean good practice scores in the 13 participating Baltic transition economy cases are the highest in this comparison, slightly outscoring both the Nordic as well as the West European LA21 processes. In fact, the Baltic cities participating in the project scored best of all participants. This means that the scope and content of the ongoing LA21 process in the Baltic area was, on average, at a very high level in 2001.

There seems to be an obvious difference between the two reference groups. It is evident that despite the low overall number of LA21 processes in the Baltic transition economy countries, the scope and content of the existing processes are on a level that is comparable with the forerunner LA21 processes.

In contrast, the LA21 process qualities of the other 17 transition economy cities are clearly lower (mean value of 40.9), close to the figures of the participating Southern European cases.

CONCLUSIONS: EXPLAINING THE DIFFERENCE

This brief pilot study shows that (at least regarding the scope and content of the LA21 processes) the Baltic area transition economy cases are far ahead in comparison with other transition economy countries. I am naturally aware of the limitations of this material; questions can be raised regarding the validity and reliability of the material. The sample for the part of the study that measures the scope and content of the processes can hardly be called independent. However, since the total number of LA21 processes is rather low in at least both groups of transition economy countries, the number of LASALA cases counts for about 10 per cent of all cases – and to some extent this eliminates problems with the sample. The influence of the active Nordic region seems to be obvious, both as an example and also as a source of diffused policy models.

For the activity level variable, there was a significant difference between the Nordic and Baltic levels of LA21 activity; but no significant difference between Baltic and other transition economy activity levels could be found. However, even in this case, at least the reliability of the material can be questioned. The *Second Local Agenda 21 Survey* (2002) by the ICLEI and the UN is not satisfactory for all parts – for example, several countries have not reported on LA21 at all. This part of the analysis shows, however, that the explanation by Lafferty and Eckerberg (1998) that national campaigns could be decisive for successful LA21 processes is only partially valid. There are other ways to success, at least at the local government level.

Can my other assumption explain the difference? In order to look more closely at this, it is important to analyse where the difference between the two groups can be found. Figure 7.1 provides an analysis of the different criteria and the different elements in the good practice score definition.

The clearest difference between the two groups of participating cities can be seen in the criteria of planning (see Figure 7.1). The Baltic cities are clearly ahead of the other transition economies, as well as their Nordic counterparts, in their sustainability planning, at least according to our LASALA good practice scoring. Other significant differences can be found (in favour of the Baltic cases) in the awareness criteria and the problem identification criteria. The Baltic cities also show a considerably higher score for the criteria that measure partnerships between different actors: networking structures.

Finally, this study puts its hypothesis under one more test. While collecting data for the self-evaluation, the LASALA research team also gathered other information on the LA21 processes. One of the questions was to what extent the

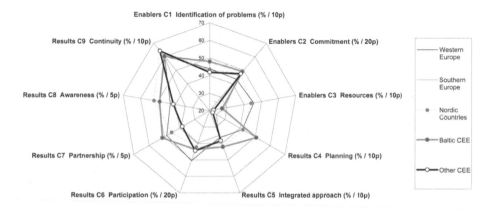

Figure 7.1 *Regional differences for the good practice criteria*

Source: LASALA database

participating LA21 processes had used models and examples from other actors on the 'sustainable development market'.

Table 7.7 lists the distribution of the mean values for this variable. It shows clearly that cases in both Baltic and other transition economy countries are, to a higher degree than Nordic or West European cases, dependent upon support from international and national networks, and, further, that national governmental support is rather weak in terms of LA21 implementation. The fact that foreign models and examples are essential for both groups is one conclusion that can be drawn from Table 7.7. However, these figures do not support the assumption that there is a visible difference regarding international multilevel governance networks. The results, in fact, lean towards a higher level of network contacts for the group of other transition economy cases. Instead, direct contacts with twin cities seem to be more important in the Baltic region, where there is stronger support from local government organizations and external research institutes and universities.

To conclude, the Baltic region countries are, to some extent, at least considering the scope and content of their best practice LA21 processes ahead of other transition economy cities. But, against this chapter's hypothesis, this cannot be explained by the existence of several cross-border multilevel governance organizations. However, since this chapter is based on a pilot study, further analysis must be conducted on this topic.

There is also a movement to abandon the concept of LA21 in favour of broader sustainable development concepts. This will gradually transform activities from a single LA21 project to everyday local authority governance projects.

Table 7.7 *Models and examples used by the participants in the LASALA project*

Models and examples from:	Country groups				
	Western Europe	Southern Europe	Nordic countries	Baltic countries	Other transition economy countries
Neighbouring local government	1.52	0.72	1.35	1.15	1.19
Other national examples	1.69	1.54	2.00	1.31	1.38
Twin cities	0.33	0.58	1.04	1.31	1.20
Other foreign examples	0.93	1.52	1.55	1.67	1.94
National or regional governments	1.12	0.89	1.48	0.83	0.86
Municipal associations	1.07	1.23	1.35	1.38	1.07
National NGOs	1.10	0.50	1.30	0.83	1.47
Regional NGOs	1.14	0.44	1.09	0.83	1.60
ICLEI	0.93	1.74	1.30	1.54	2.00
Research institutes and universities	1.10	1.48	1.39	2.00	1.33
European Sustainable Cities and Towns Campaign	0.77	1.75	1.13	1.31	1.87
Other international networks or associations	0.60	1.00	1.00	1.31	1.43

Notes: Mean values range from a scale of 0 (no: not at all) to 3 (yes: to a very high extent) (n = 113 to 122).
Source: LASALA database

NOTES

1 This chapter was presented as an article at the ECPR Joint Sessions in Edinburgh during April 2003. I am grateful to all commentators at the conference, as well as within the project team.
2 Data were collected through the international LASALA research project coordinated by ICLEI, Europa, and funded by the European Commission, DG Research, Fifth Framework Programme, 2000–2001.
3 This part of this chapter is largely based on the analytical part in Joas (2003).
4 This part of this chapter is largely based on the data and analysis in Joas (2003), with some updated evidence.
5 Enablers are:
   • C1: identifying relevant topic areas for the LA21 process;
   • C2: commitment to the process; and
   • C3: available resources.
   Result indicators are:
   • C4: an existing sustainable development plan;
   • C5: level of integrated approach;
   • C6: level of participation;
   • C7: partnership between the council and the community;
   • C8: level of public awareness;
   • C9: level of continuity.

REFERENCES

Agenda 21: Report of the United Nations Conference on Environment and Development, Rio de Janeiro, 3–14 June 1992

Andersen, M. S. and Liefferink, D. (eds) (1997) *European Environmental Policy: The Pioneers*, Manchester University Press, Manchester

Archer, C. (2003) 'Baltic security after ten years', in L. Hedegaard, and B. Lindström (eds) *The NEBI Yearbook 2003: North European and Baltic Sea Integration*, Springer Verlag, Berlin, pp261–272

Baldersheim, H. and Ståhlberg, K. (2002) 'From guided democracy to multi-level governance: Trends in central–local relations in the Nordic countries', *Local Government Studies*, vol 28, no 3, pp74–90

Björk, P., Bostedt, G. and Johansson, H. (2003) *Governance*, Studentlitteratur, Lund

Bonsdorff, E. (2006) 'Zoobenthic diversity-gradients in the Baltic Sea: Continuous postglacial succession in a stressed ecosystem', *Journal of Experimental Marine Biology and Ecology*, vol 330, pp383–391

Cameron, E. and Joas, M. (2007) 'EU support for cities towards sustainable development – an empirical study about failure or success at the local government level', Paper presented to the Tenth EUSA Biennial Conference, Montreal, www.unc.edu/eucc/eusa2007/

Dahlgren, K. and Eckerberg, K. (2005) *Status för Lokal Agenda 21 – en enkätundersökning 2004*, Hållbarhetsrådet, Umeå

Eckerberg, K. and Joas, M. (2004) 'Multi-level environmental governance: A concept under stress?', *Local Environment*, vol 9, no 5, pp405–412

Eckerberg, K, Coenen, F. and Lafferty W. M. (1999) 'The status of LA21 in Europe: A comparative overview', in Lafferty, W. M. (ed) *Implementing LA21 in Europe: New Initiatives for Sustainable Communities*, ProSus, Oslo

Evans, B. and Theobald, K. (ed) (2001) *Accelerating Local Sustainability – Evaluating European Local Agenda 21 Processes, Volume 1: Report of the LASALA Project Team*, International Council for Local Environmental Initiatives, Freiburg

Evans, B., Joas, M., Sundback, S. and Theobald, K. (2005) *Governing Sustainable Cities*, Earthscan, London

Goss, S. (2001) *Making Local Governance Work: Networks, Relationships and the Management of Change*, Palgrave, Basingstoke

Grönholm, B. and Joas, M. (1999) *Local Environmental Activities Within and Across Borders: Union of the Baltic Cities Local Agenda 21 Survey 1998*, Åbo Akademi University, Åbo

Grönholm, B., Joas, M. and Måtar, T. (eds) (2001) *Accelerating Local Sustainability – Evaluating European Local Agenda 21 Processes: Volume II, Identification of Good LA21 Processes. Report of the LASALA Project Team on Good Practice Selection and Analysis*, International Council for Local Environmental Initiatives, Freiburg

Holm, J., Bjørnæs, T., Eckerberg, K. and Joas, M. (2005) 'Local Agenda 21 governance in Scandinavia – modernisation and transition', Paper presented to the 7th Nordic Environmental Social Science Research Conference, Gothenburg University, 15–17 June 2005, Workshop 6: Local Governments in a New Environment, Gothenburg

Joas, M. (ed) (2000) *Local Agenda 21 – Models and Effects: An Analysis of LA21 Activities in Finland and the Baltic Sea Region*, Åbo Akademi University, Åbo

Joas, M. (2001) 'Reflexive Modernisation of the Environmental Administration in Finland', PhD thesis, Åbo Akademi University Press, Åbo

Joas, M. (2003) 'Local Agenda 21 in the Baltic Sea area: Ecological, economic and political stability for local level sustainable development', in L. Hedegaard and B. Lindström (eds) *The NEBI Yearbook 2003: North European and Baltic Sea Integration*, Springer Verlag, Berlin, pp111–126

Joas, M. and Grönholm, B. (1999) 'Participatory action and environmental sustainability – Local Agenda 21 in the Baltic Sea Region', *Finnish Local Government Studies (KTA)*, no 3, pp250–263

Joas, M. and Grönholm, B. (2004) 'A comparative perspective on self-assessment of Local Agenda 21 in European cities', *Boreal Environmental Research*, vol 9, pp499–507

Joas, M., Kern, K. and Sandberg, S. (2007) 'Actors and arenas in hybrid networks – implications for environmental policymaking in the Baltic Sea Region', *Ambio*, Special Issue no 2–3, pp237–242

Jordan, A., Wurzel, R. K. W. and Zito, A. R. (2003) 'Comparative conclusions – "New" environmental policy instruments: An evolution or a revolution in environmental policy?', *Environmental Politics,* vol 12, no 1, pp201–224

Kern, K. and Löffelsend, T. (2004) 'Sustainable development in the Baltic Sea Region: Governance beyond the nation state', *Local Environment*, vol 9, no 5, pp451–467

King, V. (2003) 'What kind of civil service? Trends in public administration reform in Eastern Europe', in L. Hedegaard and B. Lindström (eds) *The NEBI Yearbook 2003: North European and Baltic Sea Integration*, Springer Verlag, Berlin, pp55–66

Kooiman, J. (2003), *Governing as Governance*, Sage, London

Lafferty, W. M and Eckerberg, K. (ed) (1998) *From the Earth Summit to Local Agenda 21: Working Towards Sustainable Development*, Earthscan, London

Lindström, Å. and Grönholm, B. (2002) *Progress and Trends in Local Agenda 21: Work within UBC Cities Union of the Baltic Cities Local Agenda 21 Survey 2001*, Åbo Akademi University, Åbo

Local Agenda 21 Survey (1997) International Council for Local Environmental Initiatives (ICLEI) and UN Department for Policy Coordination and Sustainable Development (UNDPCSD), February 1997, www.iclei.org/la21/

National Implementation of Agenda 21: A Report (2002) Department of Economic and Social Affairs, United Nations, New York, August, www.un.org

Osborne, D. and Gaebler, T. (1993) *Reinventing Government: How the Entrepreneurial Spirit is Transforming the Public Sector*, Plume, New York

Papadakis, E. (1998) *Historical Dictionary of the Green Movement*, Scarecrow Press, London

Pierre, J. (ed) (2000a) *Debating Governance, Authority, Steering and Democracy*, Oxford University Press, Oxford

Pierre, J. (2000b) 'Introduction: Understanding governance', in J. Pierre (ed) (2000) *Debating Governance, Authority, Steering and Democracy*, Oxford University Press, Oxford

Second Local Agenda 21 Survey (2002) Background Paper no 15, Department of Economic and Social Affairs, Commission on Sustainable Development and International Council for Local Environmental Initiatives, www.iclei.org/la21

Internet sites

Alakivi, I. (2007) 'Efforts and progress of Estonian cities, municipalities and counties to implement sustainability principles and accelerate sustainability', Abstract to Seville 2007, www.ell.ee/failid/Sevilla2007/Abstract_C03_Irja%20Alakivi_Estonia_Template_070306_final.pdf

Association of Finnish Local and Regional Authorities, www.kunnat.net

Staðardagskrá 21 of the Association of Local Authorities in Iceland, www.samband.is/dagskra21

Sustainable Estonia 21, Estonian National Strategy on Sustainable Development, www.envir.ee/orb.aw/class=file/action=preview/id=166311/SE21_eng_web.pdf

United Nations Agenda 21, National reporting, www.un.org/esa/agenda21/

Short interviews/surveys by email, autumn 2001

Audrone Alijosiute, ECAT-Lithuania

Ille Allsaar, Association of Estonian Cities

Marita Nikmane, Daugava's Fund Ltd, Latvia

Union of the Baltic Cities, Head of Environment Commission Björn Grönholm (winter 2007)

Union of Russian Cities through Vladimir Volkov, Korolev Town Administration

Part IV

Interactions between Regional Governance for Conservation Policy and the Use of Local Livelihood

8

Regional Actors Caught between Local Livelihood and International Conservation: Grey Seal Conservation Controversy

Riku Varjopuro and Aija Kettunen

INTRODUCTION

Nature conservation can cause conflicts when there are incompatible interests in environmental resources. This especially happens when conservation infringes on local economic activities. This chapter discusses how the successful conservation of grey seals has caused economic losses to coastal fishing. Conflict has arisen because seals take fish from nets, which leads to substantial economic losses to fishermen. This case is especially complicated and unusual because the problem has arisen as a side effect of successful conservation, which contrasts with many other nature conservation conflicts where the relationship between economic activities and conservation is direct, and where economic activities jeopardize conservation goals.

Many papers have been written about local nature conservation conflicts, outlining how a conflict is experienced and perceived in one particular region, and describing the activities that the conflict provokes (e.g. Harrison and Burgess, 1994; Agrawal and Gibson, 1999; Brechin et al, 2003). This chapter covers much of the same ground; but our main aim is to relate regional-level activities to a larger context, including policies and actors on multiple levels, because the case we are studying cannot be fully understood without looking at the various levels simultaneously. In fact, this perspective is often neglected in the analysis of conflicts involving natural resources (Agrawal, 2003). Indeed, the multilevel nature of the case is captured both by the decision-making structure and by the specific nature of the problem itself. Thus, whereas important decisions about the conservation of the grey seal in Finland are initially made at the national level, enforcement is subsequently delegated to regional levels. National decisions are, however, based on international laws and conventions, such as the European Union (EU) Habitats Directive and the Helsinki Commission on the Protection of the Marine Environment of the Baltic Sea Area's (HELCOM) recommendation for the protection of seals in the Baltic Sea. This multilevel nature is further emphasized here, where successful conservation of the grey seal causes problems for livelihoods at both regional and local levels.

In the final section of this chapter we explore the implications of this case study on the wider governance discourse. Scholars have suggested that the transition from government to governance is needed when the ability of central government to steer society has weakened. For example, global and transnational environmental problems, globalization and changes in policy problems demand new types of governance (Jordan et al, 2003). We ask in what ways the studied case helps us to understand what that change might mean in concrete terms. In order to answer this question, we need first to plunge deep into the case, since it is in the actualities of dealing with this complex subject that the transition from government to governance manifests itself, if it is to manifest itself at all.

THE CASE STUDY

The methodological approach in this chapter is a descriptive one. By means of empirical data, we describe what has happened in the study area – namely, the Kvarken area in Finland and Sweden (see Figure 8.1). To begin with, we outline how the present situation emerged (i.e. how the grey seal policies and the fishery developed). After describing the processes that led to the conflict, we concentrate on a Kvarken Council project on grey seals, which created a new regional network to manage the problem. In our analysis, we describe the regional actors' own experiences of the project. Finally, the case analysis is related to a more general discussion on governance of complex multilevel issues.

The Kvarken region's actors cooperated to organize a project called *Gråsälen i Kvarken* (Grey Seal in Kvarken, or GiK) that was conducted between 2001 and 2003. The project was managed by Kvarken Council, which is a cross-border cooperation association for the Ostrobothnian counties in Finland, and the counties of Västerbotten and Örnsköldsvik in Sweden (Kvarken Council, 2000). Kvarken Council worked together with eight active partners in the project: four from Finland and four from Sweden. The partners were regional fisheries organizations, hunters' organizations, regional nature conservation authorities, and the regional fisheries authorities from both countries. In addition to these actors, several other groups were informed about the activities and were invited to seminars, giving them an opportunity to comment on the project as it progressed.

Two kinds of empirical material – namely, primary and complementary – were collected for this study. The primary material is made up of qualitative

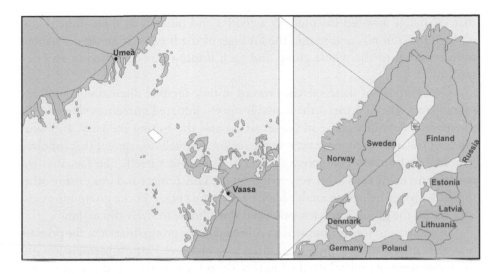

Figure 8.1 *The study area Kvarken, located in the Gulf of Bothnia,*
Northern Baltic Sea

Note: Counties bordering the narrow sea areas on both sides (i.e. in Finland and Sweden) form the Kvarken. The main towns and county capitals are Vaasa in Finland and Umeå in Sweden. The white box represents a grey seal reserve on the Finnish side.

Source: Riku Varjopuro

data collected in a group interview and also in four individual interviews of those people actively involved in the project. The group interview was conducted during October 2004. All of the active partners and especially those directly involved were invited to discuss the project in a meeting organized by one of the authors (Varjopuro). All partners, with the exception of the nature conservation authority from Sweden, attended the meeting. In fact, as the discussion revealed, this partner was not as active during the project as the other organizations were. The group interview served the purpose of collecting qualitative data (e.g. Morgan, 1996). The meeting took place in the premises of Kvarken Council in Vaasa, Finland, in the very same room where the GiK project itself had many of its meetings – a familiar place selected to encourage a relaxed and comfortable feeling for the interviewees (on the importance of interview settings, see, for example, Hammersley and Atkinson, 1995).

The interview had two parts. First, a discussion on the history of the project was initiated, in which the group produced a shared view of the background and development of the project (see Schusler et al, 2003). This 'shared history' produced a common understanding of the important issues that led to the establishment of the project and described its various phases. The second part of the interview involved a self-assessment. The group was divided in two smaller groups that separately assessed the project's activities and impacts in three dimensions: ecological, economic and social. The findings of the break-out group discussions were reported to the whole group and each finding was discussed in an open session.

One purpose of these exercises was to initiate focused discussion among the participants. The first part – the shared history – initiated numerous views on the controversy between grey seal conservation and the fishing industry, and what kinds of reactions there have been in the region. The self-assessment that followed concentrated on more concrete aspects of the project itself. The fact that the participants knew each other very well from the GiK project and from many other shared activities helped to establish a friendly atmosphere in the group interview. As a result, the group interview produced three hours of lively discussion.

In addition to the qualitative data collected in the group interview, the primary material consists of four individual interviews of active Finnish partners in GiK: the fishermen's organization; the hunters' organization; the nature conservation authority; and the project manager from Kvarken Council. These interviews were conducted in 2003 and 2004 prior to the group interview. In addition, complementary material is provided for the purpose of giving a better understanding of the project and its relation to a larger context. This complementary material consists of the documents produced by GiK and 12 interviews of individuals representing groups or organizations interested in the seal–fishery interaction. Those interviewed were fishermen, hunters, environmentalists and nature conservation authorities in the region, and hunting and fishing authorities at the national level.

BACKGROUND: GREY SEAL POPULATIONS AND INTERACTION BETWEEN SEALS AND COASTAL FISHERY

In order to understand the current situation and its dynamics, it is important to know how the situation evolved. Our first analysis looks at the changes in the grey seal population. This is followed by a description of the development of grey seal conservation. Finally, we discuss the development of coastal fishery in the region and the implication of geographical differences.

Figure 8.2 presents changes in the grey seal population in the Baltic Sea from 1900 onwards. It has been estimated that in 1906 the number of grey seals in the Baltic Sea comprised 88,000 to 100,000 individuals (Harding and Härkönen, 1999); but the population declined at different rates throughout most of the 20th century. The trend was reversed during the 1980s, and the grey seal population has been growing rapidly ever since. The current estimate of the population size is above 20,000 (Halkka et al, 2005). The growth of the seal population has especially taken place in the northern parts of the Baltic Sea. For instance, in the Finnish coastal areas the (counted) number of grey seals increased from 500 to

Figure 8.2 *Estimated size of the grey seal population from 1900 onwards*

Source: Harding and Härkönen (1999); Halkka et al (2005)

600 during the late 1980s, to 1900 to 2200 during the late 1990s (Below and Soikkeli, 2000).

Seals and fisherman are competitors in the sense that they both aim to catch their share of fish resources. The number of grey seals is naturally a relevant factor in determining the intensity of competition, which occurs at the population and individual level. At the population level, seals and fishermen compete over fish stocks, whereas competition at the individual level takes place in fishing nets when seals take fish from the nets or break the nets. Currently the stakeholders are especially concerned about competition at the individual level.

Seal conservation policy is relevant to the relationship between seals and fishery: conservation policy is highly focused on grey seal numbers, and the number of seals influences the intensity of competition between seals and fishermen. Another reason is that conservation of grey seals partly determines the actions that are allowed to minimize losses to the fishing industry. For instance, grey seal conservation policy stipulates how many seals can be hunted.

Between the 1880s and 1975, the only major change in Finnish legislation regarding the grey seal was that in 1909 the state introduced a bounty for killing seals. This bounty gradually increased and the last increase occurred in 1964 (Kalliomäki, 1998; Ylimaunu, 2000). Until 1975, the official seal policy in Finland focused only on managing the hunting of seals; but the policy changed rapidly during the 1970s when the decline in the seal population began to worry biologists. Indeed, the decline was already observed during the 1950s and 1960s (e.g.

Bergman, 1957); but by the 1970s the need for conservation was clearly recognized (e.g. Niemelä, 1973). This coincided with an increase in general environmental awareness, which changed the way in which people and the state valued grey seals (Ylimaunu, 2000). The first restriction on hunting was introduced in 1975 and restrictions soon became increasingly stringent, until grey seal hunting was totally banned in 1982.

Finnish seal policy evolved once again when hunting was reinstated in 1997, when strong lobbying by the fishery sector finally paid off. At first, the Ministry of Agriculture and Forestry, which is the highest authority in managing all game species in Finland (i.e. species managed under the hunting legislation), granted licences to shoot 30 seals in 1998. Since then, the quota has been increased gradually, and for the hunting year of 2007/2008, 685 seals may be shot nationwide, while in the study area the regional quota is 180 individuals (Hunters' Central Organization, 2007).

Managing seal hunting is not the only concrete activity in Finnish grey seal conservation policy. Nature conservation authorities and environmentalists have also been active, for instance, in establishing grey seal reserves. The process of designating these reserves began in 1995 when the grey seal working group of World Wide Fund for Nature (WWF) Finland proposed to the Ministry of the Environment that Finland should designate grey seal reserves. The WWF's working group approached the Ministry of the Environment instead of the Ministry of Agriculture and Forestry because designation and management of all protected areas in Finland is the responsibility of environmental administration. This also applies to seal reserves. The background of the designation was HELCOM Recommendation 9/1 on the protection of seals in the Baltic Sea and the EU Habitats Directive since both of these international commitments require protection of the grey seals' resting and breeding places. Seven grey seal reserves were designated in 2001 along the Finnish coasts. One of the reserves, called Snipansgrund-Medelkalla, as shown in Figure 8.1, is located in the study area (Bäck et al, 2004; Metsähallitus, 2004).

During the summer of 2004, the Ministry of Agriculture and Forestry began preparing a national management plan for the grey seal population. This was the first attempt to create a national seal policy in a coordinated manner.

Seal policy has especially focused on population size, but changes in grey seal numbers are not the only factors determining the interaction between seals and fishing. Fishing practices, particularly selection of fishing grounds, types of nets and target species, are as important factors as the number of seals, and here we concentrate on the development of coastal fishery in the study area. In other words, we describe the development of the other 'party' to the controversy. The fishery that this chapter deals with is small scale and coastal. It has traditionally been a multi-species fishery that has used trap nets and gill nets.

During the early 1960s, coastal fishing in the study area was still restricted to inner archipelago areas. However, environmental degradation of rivers and coastal waters led to a decline of near-shore fish stocks. Consequently, fishermen had to move to new fishing grounds in the outer archipelago and to target species that were less affected by environmental degradation – namely, herring, whitefish and salmon. Technological development of nets and boats aided this change, and by the 1980s fishing in the outer archipelago and concentrating on a few species had become the dominant strategy (Tuomi-Nikula, 1981; Österbottens fiskarförbund, 1990).

Fishing during the 1990s did not experience the substantial changes that occurred in the 1970s. However, whitefish became an even more important target species because its price stayed relatively high. By contrast, during the 1990s salmon fishing became more and more regulated, which in turn led to a gradual decline in catches (SVT, 2001).

Throughout the 1990s and early 2000s, the number of fishermen in the study area gradually decreased. Today there are approximately 180 professional fishermen in the area and 500 fishermen who receive some income from fishing (Kvarken Council, 2003). Those with more than 30 per cent of income from fishing are classified as 'professional' fishermen in Finland, which is an important distinction in the context of the seal controversy: only professional fishermen are eligible for most of the state-initiated mitigation measures that are discussed below in more detail.

The change in fishing strategy between the 1960s and 1980s is significant in the context of the coastal fishery's interaction with seals: during the changeover period seals were practically absent in coastal areas. An unfortunate consequence of this was that when fishermen selected new fishing grounds and adopted new fishing technology, they did not have to take seals into account, and the decisions they made at that time meant that they would later be vulnerable to seals.

EMERGING CONTROVERSY AND INTRODUCTION OF MITIGATION MEASURES

During the 1990s, the situation developed into a real problem for fishermen. For instance, during the interviews, the year 1995 was mentioned as the point when the problem became very serious on the Finnish side of Kvarken. In one interview, the emergence of the problem was described in the following way:

> Seals did cause damage throughout the 1980s, too; but there weren't really any discussions about getting something done. Damages were not that severe. It was a problem in the open sea, not yet in coastal fishing... Fishermen tried to find their own ways. During the 1990s we calculated ... it was in 1995 when the fisheries insurance system

> *started to record seal damage as a separate category. Damage was so significant then. (Interview: Fisheries organization, Finland)*

That year the amount compensated for by the fishing gear insurance association in the Kvarken region for damage to fishing gear exceeded 100,000 Finnish markka (17,000 Euros) for the first time (Kaarto, 1998).

The seal fishery controversy was discussed in a series of meetings and seminars during the 1990s. One of the first and biggest events was a meeting in Geta, Åland Islands, in 1995. It was a joint Finnish–Swedish meeting that the Swedish project *Sälar och Fiske*[1] and Naturvårdsverket (the Swedish Environmental Protection Agency) organized so that authorities, scientists and stakeholder organizations could discuss the size of the seal population in the Baltic Sea and the HELCOM recommendation on the protection of Baltic Sea grey seals. One outcome of the meeting was the establishment of experimental seal hunting in Sweden and in Finland; but the meeting also discussed the need for designating protected areas for seals (Karttunen, 1995). Some of the Kvarken region actors attended the meeting.

In Finland, fishermen began to actively make their seal problem known during the mid 1990s. There were, for instance, special issues devoted to the 'seal problem' in the fishermen's newsletter *Kalastaja* and in the fishery sector's journal *Fiskeritidskrift för Finland* in 1998. The editorial of *Fiskeritidskrift* stated in the summer of 1998 that seal hunting must be allowed quickly 'before the number of (problem) seals becomes too high' (Nikiforow, 1998, brackets in the original).

When the severity of the problem became evident at the end of the 1990s, measures to mitigate losses were introduced. Many of the ideas for mitigating the problem originated in Sweden where the problem had already escalated during the 1980s, and where the project *Sälar och Fiske* was launched in 1992 with the aim of finding solutions. Cooperation across the border is frequent and Swedish initiatives were well known on the Finnish side. For instance, the authorities in Sweden started to pay compensation to fishermen in 1986 (Naturvårdsverket, 2001), and during the early 1990s the Kvarken region fishermen's organization appealed to the national authorities in Finland and demanded that seal damage should be compensated. During the same time period, fishermen approached game managers in Finland and Sweden to request the resumption of seal hunting. Neither of the initiatives received an immediate positive response.

Currently, there are several mitigation measures available. Seal hunting, which resumed in 1998, was the first mitigation measure that was introduced in Finland. In addition, there are three economic instruments available in Finland to ease the seal problem: two are used to mitigate economic losses that seals cause to fishing, and the third subsidizes the adoption of technical mitigation measures.

In 2003, one of the compensation schemes was paid by the fishery authority to professional fishermen who suffered over a 20 per cent loss of catch due to seal predation during 2000 and 2001. The damage was compensated for according to an average level of losses in different fishing locations and with different fishing methods. The compensation scheme was temporary, as stipulated by European Union state aid regulations (European Commission, 2002; Similä et al, 2006).

Professional fishermen also have another opportunity to receive compensation through an insurance system that covers all physical damage to nets and other equipment, including damage caused by seals. The insurance system was established during the 1930s and is partly financed by insurance fees and by the state (see Österbottens fiskeriförsäkringsförening, 2002).

Funds are also available for developing and investing in fishing gear that is less vulnerable to seal damage, which is the third economic instrument. A recent change in the Fishing Decree made so-called 'seal-proof trap nets' eligible for subsidies from Financial Instrument for Fisheries Guidance (FIFG) funds, which is an EU structural fund. In the Kvarken region, over 20 fishermen have applied for these subsidies, with most applying for two sets of gear. Nationwide, 90 fishermen submitted applications for subsidies in order to purchase 250 trap nets. They will be subsidized only once, and the subsidy amounts to 70 per cent of the value of the first two sets of gear and 50 per cent of the rest (MoAF, 2005).

Despite the many relevant instruments available, interviews in the region and documents that the fishery administration has published (e.g. MoAF, 2002) indicate that the problem is very severe. Indeed, it has been mentioned as the biggest single problem for coastal fishery in Finland. For instance, during 2000 and 2001, the estimated loss of catch in salmon and whitefish fishing in the study area varied between different locations and gear types, from 0 to 80 per cent, while on average the loss was 46 per cent (TE-keskus, 2002).

It seems evident that none of the practised measures, when applied in isolation, is effective in reducing the damage, and that together they do not form a sufficiently effective combination of measures. National authorities have introduced mitigation measures; but coordination between different measures and between actors in different sectors has not been very effective so far. However, in 2004, the Ministry of Agriculture and Forestry started to prepare a national management plan for grey seals, which aims to coordinate activities. The national plan was published in 2007 and its effects can be expected in coming years. So far, Kvarken Council's grey seal project has been the only coordinated attempt in Finland to manage the conflict. Below, we concentrate on that voluntary and collaborative attempt.

Table 8.1 summarizes the development of grey seal conservation in Finland and the important changes in the Kvarken fishery described earlier.

KVARKEN COUNCIL'S GREY SEAL PROJECT

This section concentrates on the regional actors' own experiences and descriptions of their collaborative project, which aimed to find solutions to the grey seal controversy in the region.

Table 8.1 *Key events in the history of seal conservation and in the region's fishery*

1970s	During the decade:
	• Fishing moves from inshore to the outer archipelago in Kvarken and the number of fishermen decreases.
	• Whitefish and salmon catches increase.
	• 1975: Springtime (10 March–31 May) is closed season for seal hunting.
	• 1976: This marks the end of bounty hunting.
1980s	During the decade:
	• Fishing concentrates more and more on salmon and whitefish in Kvarken; salmon prices dwindle, while whitefish prices increase.
	• 1982: There is a total ban on grey seal hunting.
	• 1988: HELCOM recommendation on seal conservation (ban on hunting and the designation of reserves).
1990s	During the decade:
	• Damage caused by seals intensifies in Kvarken; in 1995, insurance compensation exceeds 100,000 Finnish markka (17,000 Euros) for the first time.
	• Whitefish prices increase.
	• The number of fishermen declines in Kvarken.
	• 1995: WWF Finland (grey seal working group) proposes the designation of grey seal reserves to the Ministry of the Environment.
	• 1995: The seal seminar takes place in Geta, Åland Islands.
	• 1995: Finland joins the EU (Habitats Directive).
	• 1997: Experimental hunting of seals accepted to test the possibility of reducing damage to the fishery in Finland and Sweden.
	• 1998: Hunting is allowed with a small quota in Finland.
2000s	During the decade:
	• Hunting quota on grey seals increases gradually and the hunting season extends.
	• Seal damage to fishing gear intensifies.
	• Whitefish becomes the most important species commercially.
	• Testing and development of new trap nets to avoid seal damage occur.
	• 2001: Designation of seven grey seal reserves in Finland (by the Ministry of the Environment as a response to the Habitats Directive, the HELCOM recommendation and WWF Finland's proposal).
	• 2001: Kvarken Council's grey seal project is established.
	• 2002: The EU Commission rejects the Finnish plan of a permanent compensation scheme, but accepts a temporary arrangement to compensate loss of catch in 2000 and 2001.
	• 2003: The Kvarken Council's grey seal project's action plan is published.
	• 2004: Change in fishing decree makes 'seal-proof' trap net eligible for subsidies from FIFG funds.
	• 2004: Preparation of a national management plan for the grey seal.

Background and preparation

The initiative to launch the grey seal project came from the regional fisheries organizations in Finland and Sweden as a result of many discussions and several unsuccessful attempts to influence national authorities and policy-makers. In fact, the central authorities' perceived passivity was one of the reasons why regional actors began planning their own activities. It was stated in the group interview that:

> *The project has really been needed for a long time; but national authorities have not been interested in such activities. The highest-level fisheries authorities in Finland have all along been against this kind of project. They are really reluctant to touch these issues. (Fisheries Authority, Finland)*

The seal fishery controversy was discussed in a series of meetings and seminars during the late 1990s in Finland and Sweden. In 1995, a seminar took place in the Åland Islands; but there were also three seminars in the Kvarken region. The first took place in Umeå in Sweden in 1998, where the idea of a joint project under Kvarken Council was proposed. After the seminar, the regional fishermen's organization in Finland took the initiative to Kvarken Council in autumn 1998. The response was positive, and in order to enhance preparation of the project, Kvarken Council organized the second seminar in Vaasa, Finland, in 1999. From the beginning, the idea was to bring different interest groups together to find solutions:

> *In the seminar in Vaasa, all participants were enthusiastic about cooperation in this issue ... even the environmental sector, although in the actual project some of them were not so much involved because the biological research of seals that they preferred was not included in the project. (Nature Conservation Authority, Finland)*

The third seminar was also held in Umeå, Sweden, in 2000. This seminar was organized under the auspices of the Royal Swedish Academy of Agriculture and Forestry, which 'gave even more legitimacy for our plans' as it was cited in the group interview. As a result of the seminars and discussions between partners, the project plan was submitted to Kvarken Council during spring 2000, and the project started in 2001.

Kvarken Council was an appropriate forum for cooperation since a great deal of cooperation between the actors either related to fisheries or to environmental issues already existed. Kvarken Council was also suitable because of its spatial scale and its regional emphasis. In other words, it provided regional actors with a common forum to deal with regionally relevant issues. 'Regionality' is, in fact, an important issue in Kvarken, which is characterized by a strong, shared 'Kvarken identity' that unites both sides of Kvarken across the national border. The importance of this identity was very clear in the interviews as well. Geographical closeness and a common language, namely Swedish,[2] are issues that connect people on both sides of Kvarken:

> *Thanks to our good contacts ... and our proximity to the Swedish side, we cooperate a lot... We are all Swedish speaking and have close contacts. For me these are [important issues] that helped to create the project. (Fisheries Authority, Finland)*

> *This is regional management [of seals] and we speak the same language. (Hunting organization, Sweden)*

The regional actors perceive that the problem itself is the same across the region. In particular, the spatial extent of the seals' territory was emphasized as being compatible with the regional approach of the project. Actors thus perceived that Kvarken is both culturally and biologically an appropriate area for dealing with the problem. As a consequence, partners of the GiK favour regional management. One further issue related to the question of scale is the friction between regional and national administrations – which further underlines the actors' regional identity:

> *Fisheries Authority, Finland: It is easier to work with Kvarken Council than Helsinki (capital city).*

> *Hunting organization, Sweden (responds to the previous speaker): We are alone in Västerbotten [Sweden] in the same way.*

Since the content of the project was initially discussed, the intention was to include it within a larger frame of environmental cooperation in the Kvarken region – namely, *Grön Bro* (Green Bridge) cooperation, which is one of the main forms of cooperation in Kvarken Council. However, fishery- and hunting-sector representatives wanted the project to concentrate on promoting hunting and on developing technical solutions to minimize the damage to the fishery. When the proposal was finally formulated, the environmental sector's perspectives were not given as high an importance as nature conservation authorities and environmentalists wished. It was then decided that the project was not to be part of Kvarken Council's environmental body:

> *Kvarken Council has the Grön Bro cooperation. At first we thought that the seal project would go into that; but the environmental sector did not want it there. It did not fit the image because of the emphasis on*

hunting... In fact, we could progress more smoothly from the fisheries-sector perspective when [the project] was separated from environmental issues... It did make the Regional Environmental Centre [in Finland] cautious, though. (Fisheries organization, Finland)

In this sense, even though the project partners greatly emphasized the close cooperation between all interests groups in the region, it seems that not all interests had equal weight in the project. However, the project partners themselves shared the view that all were welcome:

I think that [all interest groups] could have come to the project if they had chosen to come. For instance, environmentalists could surely have come and the Regional Environmental Centre... All were invited. (Hunters' organization, Finland)

In fact, the project partners were disappointed that environmentalists and environmental authorities were not keen on participating in the project:

In 2001 we had a meeting in this room with nature conservationists. It was four or five groups here ... all the relevant [groups] from Finland. Most of them were not interested in the project. They had nothing against hunting; but they were not ready to get engaged either... In Sweden, we had a similar meeting. Only two people from environmental groups came. They too had nothing against the project, but didn't want to participate in the project either. (Fisheries Authority, Sweden)

This contrasts with the experience of one of the region's environmental activists, who did not feel very welcome:

I was invited to join the project team; but after I had criticized some of the views, soon no invitations or other material was sent to me. I criticized seeing seals as a renewable resource before the necessary scientific studies had been done. (Environmentalist, Finland)

Project activities

The project was funded by the EU's INTERREG IIIA Kvarken-MittSkandia programme. Co-financing came from Kvarken Council, the Nordic Council of Ministers, the County Administration of Västerbotten in Sweden, and the Employment and Economic Development Centre of Ostrobothnia (Fisheries Unit) in Finland.

The main objectives for the project were to:

- find a common understanding about the seals' role in the Kvarken region's ecology;
- reduce the damage to the fishery industry by developing appropriate fishing gear; and
- perceive grey seals as a renewable and utilizable resource, the use of which can be an income-generating activity (Kvarken Council, 2003).

The project placed a lot of emphasis on the first objective – on finding common understanding. According to interviews of GiK participants as well as other actors in the region, this was important because finding a common understanding of the situation would help discussions between stakeholders in the project. In some interviews, this was mentioned as an exercise that pulled different and even previously antagonistic groups closer together.

Other project activities concentrated on informing the public and national- and even international-level actors about the problem; on fishing gear development; on training hunters; and on developing products from seal meat, skin and fat. The project published a regional action plan to guide activities related to the seal question (Kvarken Council, 2003). The plan suggested that actions be taken in the region and at national and international levels.

During the group interview conducted with GiK partners in 2004, participants assessed their own project. They discussed concrete project activities and other available measures to minimize damage in two groups. Activities discussed were as follows:

- development of 'seal-proof' fishing gear;
- training of hunters;
- development of seal products;
- dissemination of information about the seal problem and the project's activities; and
- cooperation and preparation of the region's action plan.

Participants also discussed other available measures to minimize the conflict. These included seal hunting, compensation, financial support of 'seal-proof' fishing gear and fisheries insurance.

Assessment results were very positive, and in this sense they may reflect the wishes of project partners, rather than the real outcomes of the GiK project. In fact, one of the project partners made this observation in the group interview: he would have preferred more critical assessments of their own projects. Even though the results of the self-assessment may be uncritical, the discussion during the group interview provides interesting material about the issues that project partners perceive as being important. In the remainder of this section, we concentrate on these issues.

One of the main themes in the discussion was that most of the project activities improve the viability of coastal fishing either by reducing damage or by compensating for it. As mentioned above, the seal problem is perceived to be a threat to the future of the whole coastal fishery, and this was originally one of the main motives for the project. Another important theme in the discussion was that grey seals should be perceived as a renewable resource, rather than merely as an animal that causes damage. Project partners felt that, in this respect, the project was successful, especially when compared to other coastal regions where fishermen perceive seals only as a pest:

> We had a meeting of fisheries organizations [in Sweden] last week. I found it very astonishing when people talked about seals: it was only a pest ... only to shoot them. No nuances, nothing of [seals as] a resource or exploitation [of seals]. (Fisheries organization, Sweden)

The GiK partners linked perceiving seals as a renewable resource and the survival of coastal fisheries to a larger discourse of *Levande skärgård* (the living archipelago), which is a common discourse in Finnish and Swedish coastal areas (see, for example, Peuhkuri, 2004; Carlberg et al, 2005).

The third theme that rose from the self-assessment centred on the conflict between stakeholders, which some of the activities were designed to reduce. For instance, the project succeeded in getting different interest groups together, and this lessened conflict in the region. Legitimizing hunting, rendering it more professional and controlling culls also helped to ease the conflict because skilful hunting means fewer wounded seals and less painful death for seals, which are important issues for nature conservationists. However, hunting and using seals as a resource were also seen as a potential source of conflict, although informing people about the plight of fishermen and about project activities was thought to make people understand and accept the GiK project's goal:

> It is important to sell the whole way of thinking, the whole 'living archipelago' concept, not just the [seal] products. (Kvarken Council, Finland)

'Gråsälen i Kvarken' between international policies and local livelihood

Building a common understanding in the area was one of the main objectives of the project. To achieve such a goal requires good relationships between actors in the region. As we pointed out at the beginning of this chapter, the controversy has a strong multilevel nature, from international to local levels. Therefore, to really make a difference, the region's actors should be able to extend their activities to other levels. This section examines the GiK and its activities in relation to different levels: regional, higher policy-making and local levels.

Regional-level actors have recognized that current relationships between interest groups are better than a few years ago. Even actors not directly involved in the project, such as the Regional Environmental Centre in Finland, admit that GiK activities have contributed substantially to this development. Regional consensus seems to be relatively strong at the moment, although it was reached through compromise. Actors in the region had interests and duties that had, to a certain extent, to be fulfilled; but all parties have also been able to listen to other groups:

> *Author: Was the process of preparing the regional action plan easy?*
>
> *Fisheries organization, Finland: Well, we all had to be flexible; none of us got our own views directly [into the plan]. We worked very carefully on it for six months. There weren't big problems; but we thought carefully and talked about individual sentences… Potential collision of hunting and nature conservation issues was discussed a lot. Fisheries groups had to make many concessions.*

The GiK thus succeeded in mediating between regional-level actors; but the project was, in many respects, also directed at higher levels. For example, resources such as hunting licences or the EU's FIFG funds that were granted from higher levels were made relevant to the region's needs. At the regional level, the main actors (especially the authorities) had some autonomy in deciding about the use of available resources. Through the project, these resources were channelled to serve the objectives in the region. A good example of this is how hunting licences were granted in the region. The Ministry of Agriculture and Forestry decides on the seal hunting quota; this is further divided into regional quotas. The regional quota is thereafter distributed to hunters in the form of hunting licences. In June 2003, there were still two seal hunting licences left from the regional quota and the hunting season was open until 31 July. These licences were not granted earlier because the regional game management authority had reserved these especially for killing seals that were known to regularly visit trap nets, and they could be used most effectively for this purpose during the summer when fishing is active.

Another example of the GiK creating a bridge to higher levels is the way in which the project aimed to influence national and international administration or policies. The GiK project had, in fact, some national influence. First, it served as a model to other coastal regions trying to find solutions to seal damage:

> *Now we have other seal projects in Finland. All of these have taken us as a model. I have sent our project plan, so that they can see how we built the project. There are [seal] hunting training courses and 'seal-proof' trap nets are being tested. (Fishery organization, Finland)*

Second, during the summer of 2004, the Ministry of Agriculture and Forestry started preparing a national management plan for grey seals. Kvarken Council was invited to join the group preparing the plan. In this sense, the GiK project has become officially recognized and has quite effectively crossed the border between regional and national authorities:

> *Hunters' organization, Finland: Requirement to prepare management plans [for large predators] comes from the EU. Finland must prepare them. Our project has influenced the process [the national preparation]: it has opened eyes to the problem.*

> *Fishery organization, Finland: [GiK] has built competence in this area.*

> *Hunters' organization, Finland: Yes, exactly. It means that we in the Kvarken region are trusted when it comes to seals ... that we have gathered knowledge thanks to the project.*

Third, regional actors also had activities purposely directed at the national and even international levels. For instance, leaflets were used to publicize the project, and the action plan that the project published in 2003 (Kvarken Council, 2003) included several proposed changes to national- and international-level laws and policies. The action plan suggested, for instance, that the validity of HELCOM's recommendation should be re-evaluated since the grey seal population was rapidly increasing, and that management of the population in the Baltic Sea should be handled at the regional level. A seminar, organized in autumn 2003, was crucial in improving the visibility of the project, particularly with other organizations and levels of governance. Representatives from the EU Commission, ministries responsible for nature conservation, hunting and fishing in Finland and Sweden, and environmental and fisheries organizations attended the seminar.

Participating in the GiK has increased the status of many local partners, particularly with regard to their dialogue with national partners:

> *Fisheries organization, Sweden: Our organization doesn't mean any-thing to Naturvårdsverket [Swedish Environmental Protection Agency] and Fiskeriverket [Swedish Board of Fisheries]; but when I say that we have this project, then we have weight.*

> *Hunters' organization, Sweden: I think that we have some weight because we have so many different groups in the project.*

In sum, the GiK managed to cross the borders between regional, national and even international levels in various ways.

The third angle through which to view the GiK is to look at it at the local level. The project was, of course, initially a response to local-level problems and some of the project activities, such as training hunters and developing appropriate fishing gear, were particularly relevant to the local level. In addition, local-level actors (e.g. fishermen and hunters) communicated frequently with different project partners, even though they were not present in the project meetings, and they were represented in the project by their organizations.

In interviews with fishermen and hunters, it can be observed that they had recognized the project's value in developing fishing equipment and seal-related products, and in training hunters. Particularly the activities that dealt with hunting and utilizing seals were mentioned as being particularly useful because they were perceived as revitalizing a long seal hunting tradition in the area. These activities were also valued because there is a strong belief among fishermen and hunters that seal hunting will reduce loss of fish catch and damage to equipment by seals.

However, fishermen generally argued that developing fishing gear is of little benefit since they do not believe that effective seal-proof fishing equipment can be constructed. In this respect, it can be argued that had the project planning been in the hands of fishermen, this aspect of the project would never have been pursued:

> It is good that [GiK] came with a lot of facts and they had thought about markets for seal meat and skin. They worked also a lot with 'seal-proof' fishing gear and that's bad. Seal products ... clothes ... have been fine ... the project had light and dark sides. Seal-proof gear was not good. (Fisherman, Finland)

The project had a strong emphasis on fishing gear development in spite of fishermen's scepticism, of which the project partners were well aware. Regional-level actors had resources and enough power to conduct fishing gear tests without the consent of the majority of local fishermen. However, testing the gear was done in cooperation with fishermen, with the result that those fishermen who took part had a very positive perception of the equipment. The tests proved that the new trap net gives better catches than traditional trap nets in areas where seals are abundant. The project tested the viability of the net and spread information about the gear among fishermen. This kind of activity is typical of information-based policy instruments, which often have a top-down structure of influence.

DISCUSSION AND CONCLUSIONS: GIK AND MULTILEVEL ENVIRONMENTAL GOVERNANCE

This chapter has described the controversy between the conservation of grey seals and coastal fishing, and reactions to the controversy in one coastal region: Kvarken in the Northern Baltic Sea. The basis of the current situation was laid down during the period when the number of seals was low and fishing experienced several changes. The main goal of seal conservation – making the population grow – was established during the 1980s[3] when seals were seriously threatened; consequently, the focus of conservation has been on the size of the population. The problem with fishing appeared after almost two decades of successful seal conservation, and it was not anticipated by seal conservation policy or by the support mechanisms for fisheries. This resulted in a lack of measures to control the economic losses that seals may cause and, for quite a long time, in an inability to even recognize the problem. In the study area, regional actors built their own response to the controversy and managed to mediate between different stakeholders and different levels. The controversy and reactions to it that were described in this chapter can be summarized in the following way: *international conservation policy meets local livelihood and a response to this collision is collaboration between regional-level actors. The region's activities are directed at higher and lower levels, as well as towards other regional actors.* This is described in Figure 8.3.

Scholars have suggested that the transition from government to governance is necessary since the ability of central governments to guide society has weakened. For example, global and transnational environmental problems, as well as globalization and changes in policy problems, demand new types of governance (Jordan et al, 2003). If governance and government are treated as opposite poles of a (theoretical) continuum, there are more self-governing networks of societal actors influencing government policy at the governance pole than at the state-centred government

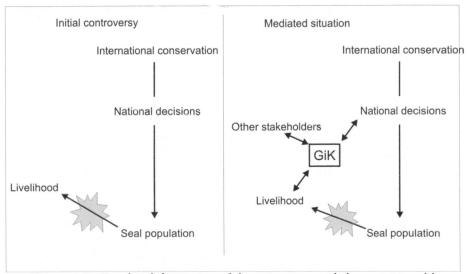

Figure 8.3 *Graphical description of the controversy and change initiated by regional mediation*

Source: Riku Varjopuro and Aija Kettunen

pole. While the government pole of the continuum is associated with more legislation, regulation, force, bureaucracy and control, the governance pole of the continuum uses more non-regulatory instruments and co-operative ways of steering, designed and implemented by non-state and state actors together (Jordan et al, 2003).

We argue that the controversy upon which this chapter is based is an example of a new kind of complex policy problem demanding new ways of management, and the Kvarken Council's grey seal project indicates clear change in this direction that is informal, voluntary and sub-national, and, at the same time, transnational (Mayntz, 1998; Paavola and Adger, 2002; Jordan et al, 2003). Mitigation measures used by Kvarken Council's project also differed from the state-centred government's style, which is regulation oriented and does not overly rely on voluntary and communicative measures (Benz and Eberlein, 1999; Jordan et al, 2003). The GiK's measures involved dialogue between stakeholders, informing the public and stakeholders at different levels, developing new fishing techniques and seal products, and training hunters. The measures were voluntary and co-operative, and were implemented by many actors. An interesting aspect of the case is that the GiK finally succeeded in having a role in preparing a national action plan for managing the grey seal populations, which further highlights that states are not the sole actors, but rather share power across levels (Fairbrass and Jordan, 2001). However, although sub-national level actors may be represented in different decision-making processes, representation is not automatically translated to influence in the same processes. Another recognizable aspect of the studied case, and which further highlights power inequalities between different actors, is that sub-national level actors are seldom autonomous because they are, to a large extent, dependent upon decision-making powers and financial resources vested in the national-level decision-making bodies (Fairbrass and Jordan, 2001). The GiK had some power; but it mainly resulted from their ability to coordinate activities in the region and to channel resources from national and EU levels to regionally relevant endeavours.

In addition to the need to approach the problem in a flexible and innovative way, the question of how to organize the governance of multilevel policy problems is also crucial. Hooghe and Marks (2003) have identified two different ways of organizing such activities that they call multilevel governance (MLG). This can be based either on stable and territorial organization, or it can be more problem driven; but they suggest that problem-driven MLG leans on a stable and territorial type, although it works differently. In the case under investigation, both of the MLG organizing types were practised. Regional authorities and interest group organizations were stable, territorial MLG actors at an intermediate level, while the GiK project constituted a flexible, interregional, problem-driven, *ad hoc* actor. In other words, the stable regional actors in Kvarken used their institutional position and resources to deal with the problem, and the *ad hoc* project capitalized on the common need for collective solution and integration of activities. The embedded nature of the problem-driven MLG compared to the more stable and institutionalized structures can be seen, for instance, in the case when the GiK project manager from Kvarken Council was invited to join an official body responsible for preparing a grey seal national management plan. As a result, the regional, informal project gained status in the established decision-making structures. The problem-driven MLG has its weaknesses, too. One key weakness is its potential blindness to 'other' views in concentrating only on the problem at hand and, furthermore, from the perspective defined by the main actors themselves. This is apparent in the Kvarken case: the regional problem-driven organization reflected the local problem – namely, seal-induced damage to fishing – but failed to take into account large-scale seal conservation goals as well as other fishery problems with equal weight. This is reflected, for instance, in the extent of participation by environmental organizations and authorities. The other weakness is its temporality. As argued earlier, the grey seal controversy cannot be completely solved, which means that it has to be managed continuously. The dilemma is that if and when the situation changes in the future, the problem must be constantly redefined through the framework of the problem-driven MLG. There is also a risk for the (so-far) successful regional cooperation. If the problem stays unsolved and even intensifies, how long will collaboration work? Will the voluntary activity of regional actors still be worthwhile in the future?

As can be seen from the previous discussion, there were a number of positive outcomes from the regional collaboration. In this case, we conclude that the transition from rigid policy structure and hierarchical government to more flexible multilevel governance explains at least part of the success of the project. The main actors were regional-level organizations and authorities, which is an important issue for two main reasons. The first advantage was that they had resources such as finances and know-how that made it possible for them to start such a project. In addition, the resources gave them confidence to highlight their region's worries and needs at national and international levels. The case therefore demonstrates that it is important that regional administration implements not only national or EU policies, but also needs to have a capacity to initiate its own programmes as well.

The second strength of regional-level actors in managing such a complex problem is that they have knowledge of the local-level problems through their links to local livelihoods. Higher-level decision-makers lack this knowledge, or have only a partial understanding of local-level situations. The ability to recognize the importance of day-to-day processes improves the capacity to discover and manage complex problems (Fairbrass and Jordan, 2001). The case clearly shows the value of the partial transfer of power from the hands of national-level decision-makers to the regions, and we conclude that such transfer should be supported politically and financially by individual states and the EU. Nonetheless, since multilevel phenomena run both ways, it is reasonable to ask how much decision-making power should be given to lower levels in issues that are by their nature multilevel phenomena? This is especially the case regarding natural resources that command large geographical and political scales: seals move in a large area (half of the Baltic Sea); migratory fish such as salmon occupy the whole Baltic Sea, etc. The risk is that regional-level economic interests will overrule the more general conservation interests. It is therefore important that managing the controversy is based on communication and power-sharing across all levels, rather than delegation of power to regional actors. However, power-sharing and efficient communication require effective mediation between the levels, which can be organized, as we have shown, in a form of regional collaboration, such as is the case with the Kvarken grey seal project.

NOTES

1   In Sweden, fishery and nature conservation sectors initiated a common project called *Sälar och Fiske* (Seals and Fishery) in 1992.
2   A majority of the population on the Finnish side of Kvarken is Swedish speaking.
3   HELCOM's recommendation on the protection of seals in the Baltic Sea was prepared during the mid 1980s, when biologists thought that grey seals could in all probability be extinct by the year 2000. Therefore, anything that could be done to save seals should be done. In this atmosphere, achieving and ensuring the growth of the grey seal population became the single most important goal of conservation.

REFERENCES

Agrawal, A. (2003) 'Sustainable governance of common–pool resources: Context, methods and politics', *Annual Review of Anthropology*, vol 32, pp243–262
Agrawal, A. and Gibson, C. C. (1999) 'Enchantment and disenchantment: The role of community in natural resource conservation', *World Development*, vol 27, pp629–649
Bäck, S., Helle, E., Lehtonen, H. and Blankett, P. (2004) 'Uhanalaiset lajit', in H. Pitkänen (ed) *Rannikko- ja avomerialueiden tila vuosituhannen vaihteessa*, Suomen ympäristö 669, Finnish Environment Institute, Helsinki
Below, A. and Soikkeli, M. (2000) 'Harmaahylje – *Halichoerus grypus*', in A. Below (ed) *Suojeluverkoston merkitys eräille nisäkäs- ja lintulajeille*, Metsähallituksen luonnonsuojelujulkaisuja A, no 121, Metsähallitus, Vantaa, Finland
Benz, A. and Eberlein, B. (1999) 'The Europeanization of regional policies: Patterns of multilevel governance', *Journal of European Public Policy*, vol 6, no 2, pp329–348
Bergman, A. (1999) 'Health condition of the Baltic grey seal (*Halichoerus grypus*) during two decades: Gynaecological health improvement but increased prevalence of colonic ulcers', *Acta Pathologica, Microbiologica et Immunologica Scandinavica*, vol 107, pp270–282
Bergman, A. and Olsson, M. (1986) 'Pathology of Baltic grey seal and ringed seal females with special reference to adrenocortical hyperplasia: Is environmental pollution the cause of a widely distributed disease syndrome?', *Finnish Game Research*, vol 44, pp47–62
Bergman, G. (1957) 'Rannikoittemme hyljekannasta', *Luonnontutkija*, vol 60, pp81–90

Brechin, S. R., Wilshusen, P. R., Forthwangler, C. L. and West, P. C. (eds) (2003) *Contested Nature: Promoting International Biodiversity with Social Justice in the Twenty-first Century*, State University of New York Press, Albany, NY

Carlberg, A., Bruckmeier, K., Elmgren, R., Frizell, B., Hill, C., Neuman, E. and Sterner, H. (eds) (2005) *Kustmiljöns framtid – Erfarenheter från forskningsprogrammet SUCOZOMA*, Rapport 2005:10, Länsstyrelsen Västra Götalands Län, Göteborg, Sweden

European Commission (2002) *Valtiontuki nro N 102/2001 – SUOMI. Hylkeiden aiheuttamien saalisvahinkojen korvaaminen kalastajille. Suullinen käsittely*, C(2002)1598fin, European Commission, Brussels

Fairbrass, J. and Jordan, A. (2001) 'Protecting biodiversity in the European Union: National barriers and European opportunities?', *Journal of European Public Policy*, vol 8, no 4, pp499–518

Halkka, A., Helle, E., Helander, B., Jüssi, I., Jüssi, M., Karlsson, O., Soikkeli, M., Stenman, O. and Verevkin, M. (2005) 'Numbers of grey seals counted in censuses in the Baltic Sea, 2000–2004', in E. Helle, O. Stenman and M. Wikman (eds) *Symposium on Biology and Management of Seals in the Baltic Area*, Kala- ja riistaraportteja 346, Finnish Game and Fisheries Research Institute, Helsinki

Hammersley, M. and Atkinson, P. (1995) *Ethnography: Principles in Practice*, Routledge, London

Harding, K. and Härkönen, T. (1999) 'Development in the Baltic grey seal (*Halichoerus grypus*) and ringed seal (*Phoca hispica*) populations during the 20th century', *Ambio*, vol 28, no 7, pp619–627

Harrison, C. M. and Burgess, J. (1994) 'Social constructions of nature: A case study of conflicts over the development of Rainham Marshes', *Transactions of the Institute of British Geographers*, vol 19, pp291–310

Helander, B. (2000) 'Havsörn och gråsälar – fortsatt ökande bestånd', in A. Tidlund (ed) *Österjö '99*, Stockholms Marina Forskningscentrum, Stockholm

HELCOM (2001) *Action Plan for the Implementation of the HELCOM Project on Seals (7th Draft)*, http://sea.helcom.fi/dps/docs/documents/Nature%20Protection%20and%20Biodiversity%20Group%20(HABITAT)/SealProject%206,%202001/Draft7_March2001.pdf, accessed 17 January 2005

Hooghe, L. and Marks, G. (2003) *Unraveling the Central State, But How? Types of Multi-level Governance*, Political Science Series 87, Institute for Advanced Studies, Vienna

Hunters' Central Organization (2007) *Hunting Statistics*, http://riistaweb.riista.fi/riistatiedot/riistatietohaku.mhtml?lang=fi, accessed 5 September 2007

Jordan, A., Rüdiger, K. W. W. and Zito, A. (2003) *Has Governance Eclipsed Government? Patterns of Environmental Instrument Selection and Use in Eight States and the EU*, CSERGE Working Paper EDM 03–15, http://www.uea.ac.uk/env/cserge/pub/wp/edm/edm_2003_15.pdf, accessed 5 September 2007

Kaasto, L. (1998) 'Merenkurkun hyljevahingot kasvavat vuosi vuodelta', *Kalastaja*, vol 4, p4

Kalliomäki, A. (1998) 'Kestävän käytön periaate luonnonvaraisten eläinten hyödynt-ämisessä. Metsästyksen ja kalastuksen oikeudellinen sääntely vuosina 1865–1998', PhD thesis, University of Tampere, Department of Public Law, Tampere, Finland

Karttunen, V. (1995) 'Hyljeongelmaa pohdittiin Ahvenanmaalla', *Suomen kalastuslehti*, vol 7, pp26–27

Kvarken Council (2000) *Kvarken – Co-operation across the Borders*, Kvarken Council, Vaasa, www.kvarken.org/banks/bank977/Kvarken_brochure.pdf, accessed 23 November 2004

Kvarken Council (2003) *Kustbefolkningen och gråsälen i Kvarken. Handlingsplan för bevarandet och utnyttjandet av gråsälbeståndet i Kvarken*, Kvarken Council, Vaasa, Finland

Mayntz, R. (1998) *New Challenges to Governance*, Jean Monnet Chair Paper RSC no 98/50, European University Institute, Florence, Italy

Metsähallitus (2004) *Grey Seal Protection Areas*, www.metsa.fi/natural/protectedareas/seals.htm, accessed 31 May 2004

MoAF (Ministry of Agriculture and Forestry) (2002) *Lohen rannikkokalastuksen kehittämistyöryhmä*, Työryhmämuistio MMM 2002:20, Ministry of Agriculture and Forestry, Helsinki

MoAF (2005) 'Valikoivien ja hylkeenkestävien rysien tukemiseen varattiin 1,6 miljoonaa euroa', Press release on 25 January 2005, Ministry of Agriculture and Forestry, Helsinki, www.mmm.fi/tiedotteet/tiedote.asp?nro=1726, accessed 9 May 2007

Morgan, D. L. (1996) 'Focus groups', *Annual Review of Sociology*, vol 22, pp129–152

Naturvårdsverket (2001) *Nationell förvaltningsplan för gråsälbeståndet i Östersjön*, Dnr 402–2695–01 Nf, Naturvårdsverket, Stockholm

Niemelä, E. (1973) 'Merialueittemme hylkeet', *Suomen luonto*, vol 6, pp249–253

Nikiforow, M. (1998) 'Sälskador växer år för år', *Fiskeritidskrift för Finland*, vol 5, p3

Österbottens fiskarförbund (1990) *Österbottens fiskarförbund 1930–1990*, Österbottens fiskarförbund, Vaasa, Finland

Österbottens fiskeriförsäkringsförening (2002) *Verksamhetsberättelse 2001*, Österbottens fiskeriförsäkringsförening, Vaasa, Finland

Paavola, J. and Adger, W. N. (2002) 'New institutional economics and the environment: conceptual foundations and policy implications', CSERGE Working Paper EDM 02–06, www.uea.ac.uk/env/cserge/pub/wp/edm/edm_2002_06.pdf, accessed 6 September 2007

Peuhkuri, T. (2004) 'Tiedon roolit ympäristökiistassa. Saaristomeren rehevöityminen ja kalankasva-tus julkisen keskustelun ja päätöksenteon kohteena', Annales universitatis Turkuensis, Sarja C, Osa 220, PhD thesis, University of Turku, Turku, Finland

Schusler, T., Decker, D. and Pfeffer, M. (2003) 'Social learning for collaborative natural resource management', *Society and Natural Resources*, vol 15, pp309–326

Similä, J., Thum, R., Varjopuro, R. and Ring, I. (2006) 'Protected species in conflict with fisheries: The interplay between European and national regulation', *Journal of European Environmental and Planning Law*, vol 3, pp432–445

SVT (2001) *Kalatalous aikasarjoina*, SVT Maa-, metsä- ja kalatalous 2001:60, Finnish Game and Fisheries Research Institute, Helsinki

TE-keskus (2002) *Keskimääräiset hyljevahinkoprosentit vahinkovuosille 2000 ja 2001*, Ostrothnia Centre for Employment and Economic Development, Vaasa, Finland

Tuomi-Nikula O. (1981) *Kalastus Pohjanmaan joissa 1800- ja 1900-luvulla*, Kokkolan vesipiiri, Kokkola, Finland

Wahlberg, M., Karlsson, O. and Lunneryd, S. G. (2004) 'Why are our grey seal population parameters different from Harding's?' Unpublished project document in FRAP Project

Ylimaunu, J. (2000) *Itämeren hylkeenpyyntikulttuurit ja ihminen-hylje -suhde*, Suomalaisen kirjallisuuden seura, Helsinki

9

Unity versus Disunity of Environmental Governance in the Baltic Sea Region

Yrjö Haila

INTRODUCTION

Because of its biogeographic history, the Baltic Sea is unique as an ecological formation. It originated as recently as 10,000 to 12,000 years ago, after the retreat of the last great glaciers from north-western Europe. The Baltic basin is relatively isolated from the main body of the Atlantic Ocean, to which it is connected through the Danish Straits. The water is brackish, with a salinity gradient of 0.6 to 0.8 per cent, to almost zero on the surface from south to north; this is a severe stress to both marine and freshwater organisms. The Baltic draws its waters from a huge and mostly densely populated catchment area that extends over eastern Scandinavia, Finland, north-western Russia, the Baltic states and western Poland; the total human population was approximately 80 million in 1989 (Jansson and Velner, 1995). This is a perfect set-up for serious pollution problems; Jansson and Velner (1995, p332) describe the particular fragility of the Baltic ecosystem as follows:

> *The large catchment area; the semi-closed cold water body, heavily stratified without tide; and a water residence time of several decades make the Baltic a very sensitive area. The intermittent pulses of oxygenated North Sea water seem to come in long intervals, leaving periods of stagnation and low oxygen levels in between.*

Public awareness about the deterioration of the Baltic environment arose during the original environmental awakening of the 1960s. A great deal of research has been done since then (e.g. Rapport, 1989; Jansson and Velner, 1995; Jansson and

Hammer, 1999; Leppäkoski et al, 1999; Gren et al, 2000; Wulff et al, 2001). The Baltic has also been a test ground – and quite successful at that – of international environmental cooperation from the early 1970s on, in the form of the Helsinki Commission on the Protection of the Marine Environment of the Baltic Sea Area (HELCOM) (see Chapter 6 in this volume). Since 2005, the Baltic has been basically an 'internal sea' of the European Union (EU), with the exception of the Russian portion at the eastern end of the Bay of Finland and around Kaliningrad. In principle, environmental governance will become more and more unified over the whole region in accordance with the guidelines of EU environmental policy.

In a formal sense, unification of environmental policy is a goal worth pursuing: unification reduces ambiguity in the criteria used in assessing environmental problems and evaluating success versus failure in addressing them, and it also helps to integrate environmental goals within other sectors of public policy. For these reasons, policy integration is a generally accepted goal in academic research of environmental policy. However, the coin has another side. The field of environmental policy is made up of a highly heterogeneous set of specific problems. At some level of resolution, differentiation of policy instruments and ways of implementation are necessary, and it is not *a priori* obvious what this level is. The general image of environmental governance ought to be formed against a realistic view of the nature of the problems it wants to address.

Meadowcroft (2002) effectively laid to rest the dream of a completely unified environmental policy, using general arguments on the multiplicity of spatial and temporal scales typical of environmental problems. This chapter's aim is to continue exploring the heterogeneity of environmental problems on a more specific ground, using the Baltic Sea as a test case. Here, problem closure is used as a specific criterion. Closure is an important notion both in scientific explanation and in politics (e.g. Dyke, 1988; Hajer 2003a, respectively). Closure is dynamic for reasons that are taken up later on; in policy analysis literature this aspect is often captured by the term framing (Laws and Rein, 2003). Operationally, we can characterize problem closure as follows: the dimensions of the problem have to be fixed in such a way that the problem and its potential solution can be identified using the same criteria. The search for closure opens up two specific questions:

1   First, what level of specificity is needed to achieve closure of specific environmental problems?
2   Second, to what extent is problem closure context specific? In other words, to what extent is a particular closure stable across variation in geographical conditions and social and political situations?

These questions can only be addressed in the light of actual specified environmental problems. Of course, we must also leave open the possibility that closure may not be achievable at all – in other words, that some problems may be patently defined within mutually incongruous, if not contradictory, frames.

A comprehensive review of the environmental problems of the Baltic region is out of the question in this context. As my primary source material, I use three recent articles published in the Finnish journal *Vesitalous* (vol 47, no 2, 2006).[1] The articles summarize results of major research projects funded by the Finnish Baltic Sea Research Programme (BIREME). Thus, we can consider them representative enough for the purposes of this analysis. However, they are almost exclusively based on the natural sciences. To complement the setting, two case studies on the social dimensions of environmental conflicts along the Finnish Baltic coast are incorporated in this chapter: seal conservation in Kvarken, the Bay of Bothnia (see Chapter 8 in this volume), and fish farming versus eutrophication in the Archipelago Sea (Peuhkuri, 2002, 2004).

In organizing and analysing this material, I make use of the notion of 'problem space', which is constructed through analogy with a physical phase space (following Garfinkel, 1981, and Dyke, 1988). The last sections of this chapter return to the question of problem closure and environmental governance.

PROFILES OF PROBLEMS

The articles in *Vesitalous* (vol 47, no 2, 2006) focus on the following problem areas: preservation of biodiversity (Leppäkoski et al, 2006), threats posed by toxins (Karjalainen et al, 2006) and eutrophication (Kuosa et al, 2006). Each of the articles spells out an overall interpretative frame and takes up a range of specific issues. In what follows, their main points are summarized. This chapter uses a relatively narrow lens to identify different issues addressed in the articles for reasons that become clear further on.

Preservation of biodiversity

Leppäkoski et al (2006) frame their article against the history of the Baltic Sea:

> *The Baltic was once an oligotrophic sea with clear water and low nutrient levels. The sea has a short history – 10,000 to 12,000 years – and is inevitably changing as it ages. Homo oeconomicus is, however, speeding up the rate of change.*

Against this background they discuss the following specific problems:

- (A1) Introduced species: all in all, about 60 to 70 alien species have established permanent populations in the Baltic. Somewhat less than 20 of them are harmful to humans; the authors name the barnacle (*Balanus improvisus*), an immigrant from North America during the 19th century, as the most harmful.

- (A2) The American mink (*Mustela vison*) is a special case among alien invaders. It has a strong presence in the Baltic archipelagos: its population originated from individuals that escaped from fur farms. Since the mink is an efficient predator, it is assumed to cause havoc in the ecological communities of the islands. A field experiment consisting of a mink removal area and a control area was established in the Archipelago Sea to test this assumption. It proved to be successful: 14 out of 22 water bird and shorebird species increased in the removal area. The experiment also showed that through predation, the mink reduces vole and frog populations on the islands.
- (A3) Eutrophication triggers a change in the composition of algal communities in the littoral zone. This change seems to be caused by reduced transparency of water, which acts as a 'forcing function' in the system, changing the relative suitability of the littoral habitat for different algal species. In particular, kelp is replaced by 'opportunistic' filamentous algae.
- (A4) The authors have experimental evidence that grazing by molluscs and crustaceans may keep abundant opportunistic species at bay locally and, hence, counteracts the impoverishment of algal communities in the littoral zone. However, they note an interesting twist to this issue: in the southern Baltic, grazing seems to become relatively less efficient when the nutrient concentration becomes higher; but in the Archipelago Sea such a transition point has not been recorded, presumably because grazer densities are higher in the north.

Toxic threats

Karjalainen et al (2006) frame their article by noting that novel substances have spread into the environment as a consequence of 'human activities,' some of them being highly toxic both to humans and other organisms. They focus on organohalogens (dioxins, furans and PCBs) and residues from oil spills. The authors discuss the following specific problems:

- (B1) Organohalogens end up in the Baltic from multiple sources: incineration of various types; industrial effluents; polluted sediments and soils; and rubbish dumps. Some types of organohalogens are used on purpose, whereas others come about as unwanted side products. Concentrations of the compounds have gone down since the 1970s, when monitoring was initiated (and appropriate analytic methods were developed); but the threat is by no means over.
- (B2) Dioxin in fish is a health risk to humans. The concentration of dioxins in fish with fatty tissues (primarily salmon and Baltic herring) exceeds the norms set by the EU; concentration increases with the age of the fish. Medical authorities consider the positive health effects of eating fish sufficient to compensate for the risk; but they recommend that salmon and Baltic herring, which is 17cm long or longer, be eaten no more than twice a month.

- (B3) As a possible method of mitigating the health risk, the authors take up the issue of intensive fishing, which results in younger fish being caught. However, their simulation models show that the effect is marginal: the risk size of Baltic herring would increase from 17cm to 17.5cm.
- (B4) As a side topic, the authors note that farmed fish fed with industrial fodder is safe for humans to eat.
- (B5) The risk of a major oil catastrophe is increasing in the Gulf of Finland as a consequence of the rapidly growing volume of crude oil transportation. This is due to the construction of new oil harbours in the eastern end of the gulf by Russian oil companies.
- (B6) Purposeful oil spills – due to washing tanks and discharging waste motor oil at high seas – are also a constant major problem.
- (B7) A new experimental research project was launched at the end of 2005 on the toxic effects of crude oil on various marine organisms in the Baltic Sea. The purpose is to identify key species at the ecosystem level that are, perhaps, particularly sensitive to oil residues in water.
- (B8) In the context of oil spills, the authors also discuss cleaning technologies. They note that cotton grass fibre, a side product of peat excavation, offers potential as an efficient and non-toxic oil absorbent.
- (B9) Several endangered species – seabirds, plants and insects – are vulnerable to coastal oil pollution. They have localized populations along the shores of the Gulf of Finland and suffer severe setbacks as a consequence of major oil spills. Protection of these species against risk calls for anticipatory measures.
- (B10) As a conclusion to the article, the authors take up a broad range of measures needed for improving the control of environmental risks: increase of knowledge (concentrations and toxicity of critical compounds; toxic effects on humans); more stringent regulations and control (the use of hazardous chemicals; the handling of waste oil; the safety of oil transportation); and new technologies (cleaning oil spills).

Eutrophication

Kuosa et al (2006) frame their article by noting that eutrophication is a basic threat to the Baltic ecosystem. Although environmental authorities have been aware of the problem for decades, the knowledge of ecosystem-level interactions and material cycles is still unsatisfactory for the needs of a sustained and cost-efficient management system. The lack of knowledge of systemic interactions in the Baltic ecosystem is the unifying theme of the article; the authors discuss the following specific aspects:

- (C1) External nutrient load discharged by rivers into the Baltic basin has a crucial effect on the eutrophication of coastal waters. The authors describe results of studies of nutrient load from the Kokemäki River catchment area (south-western Finland). The water bodies close to the coast are much more critical for the nutrient load than water basins further upriver. Lakes differ substantially from each other in terms of nutrient retention: shallow lakes with heavy nutrient concentration in the bottom sediments are particularly likely to deliver nutrients downstream.

- (C2) Biomanipulation (i.e. the removal of planktonivorous fish from lakes with accumulated nutrients in the bottom sediments) has, in some cases, proved an efficient means of reducing the nutrient load.
- (C3) Point-source pollution, particularly from St Petersburg (the source of some 50 per cent of the external load of phosphorus), is a serious problem. Efficient reduction of point-source pollution is necessary.
- (C4) Much research has been directed at the biochemical composition and reactivity of phosphorus compounds accumulated in the bottom sediments. Phosphorus occurs in sediments in several forms, which can be roughly classified as mobile and immobile fractions: mobile fractions produce internal load when they are dissolved into the water body. The internal load of phosphorus is highest at estuaries close to the coast. There are complex interactions between external and internal loads.
- (C5) Nitrogen is the limiting nutrient of algal growth in the main basin of the Baltic and also in most parts of the Gulf of Finland. Denitrification due to microbial metabolism is quite efficient in bottom sediments both in the main basin and close to the coast. However, natural denitrification is not efficient enough to balance the input from external loads. Exhaust fumes from car traffic, via aerial transportation, form an important source of nitrogen pollution.
- (C6) Ecosystem models are an essential tool for constructing scenarios of the Baltic's future state of the environment.
- (C7) Changing land use (e.g. taking arable fields out of production) in the catchment area of the Baltic is a largely unpredictable element affecting the future of the sea.

The problem space

The issues listed above form a mixed bag. This is a natural consequence of the nature of the articles: they were written to present results of broad-ranging research projects, not to define environmental problems in any formal sense. Thus, some of the specific issues raised in the articles address the same problem from different perspectives (e.g. B2 and B3; and B6, B7 and B8).

The next challenge is to identify relevant similarities and differences across this range of issues. This is done by constructing an eco-social problem space within which the problems are specified. A problem space is an analogy of a physical phase space; this analogy draws upon Garfinkel (1981) and Dyke (1988). The attribute eco-social means simply that factors pertaining to humans and the rest of nature are considered on an equal footing (Haila and Dyke, 2006).

The aim of a physical phase space is to give a representation of the possible ways of changing a particular physical system. In an analogous fashion, we can think of a problem space as summarizing the possible variations that can be found among the problems that are of interest. Problems located in the same problem space have to be, in some sense, mutually congruent. To construct a problem space, we need to identify critical dimensions along which we can characterize variation among specific problems. Such dimensions can be used as 'axes' of the space, analogously with the axes of a phase space. To qualify as axes, the dimensions need to fulfil two requirements – namely, that they are necessarily referred to when specifying a particular problem, and they can be considered separately from one another, although this does not mean 'independence' in any statistical sense.

Variation along the axes is categorical rather than numerical. In this sense, a problem space is very different from a Cartesian phase space (this idea is derived from Dyke, 1993).

So, what are the dimensions? As a starting point, we have to acknowledge that there is no 'objective' problem space: particular phenomena turn into problems only after somebody defines them as problems on various grounds and the definition receives support (Haila and Levins, 1992; Hannigan, 1995). This does not mean that the phenomena underlying problem definitions were not real: we can discard the confusing shallow 'constructivism' (following, for instance, Hacking, 2000). On the other hand, phenomena never obtain the shape of specific problems by the sheer weight of their materiality. We can also discard shallow 'objectivism' (following, for instance, Bourdieu, 1991; I come back to this point below).

In other words, the dimensions of the problem space arise through the process of problem definition. The heterogeneity of the list of topics in the previous section helps us further. We can recognize among the topics three types of characterizations that relate to problem definitions. They can be specified in the form of the following three questions:

1   First, what, precisely, is threatened/damaged (dubbed 'target' below)?
2   Second, where does the threat/damage originate from ('source' below)?
3   Third, what can be done about the threat/damage ('mitigation' below)?

These three dimensions are necessary for achieving closure of any particular environmental problem. They are therefore taken as a first approximation of the dimensions of the eco-social problem space. Table 9.1 presents categorizations of the target, source and mitigation of each of the specific problems described in the articles (Leppäkoski et al, 2006; Karjalainen et al, 2006; Kuosa et al, 2006). I have ordered the categories from more specific to more general. There is some ambiguity in this regard, particularly concerning the third axis (mitigation); but the ordering serves the purposes of investigating the structure of the problem space. Table 9.1 has attempted to be faithful to the source articles.

Table 9.1 *A categorization of the targets, sources and mitigations of the specific problems identified in the source articles (Leppäkoski et al, 2006; Karjalainen et al, 2006; Kuosa et al, 2006)*

Target
Specific economic cost: (A1; implicitly in the reference to the barnacle *Balanus improvisus*)
Specific human health hazard: (B1)
Threat to specific species: (B7)
Impoverishment of ecological communities: (A2) (A3)
Harmful effects at the ecosystem level: (B6) (C1)
Systemic damage to human use of the environment: (C1; this is not, curiously, mentioned by Kuosa et al, 2006)

Source
Purposeful human acts: (A1; partially) (B5)
Inadvertent human acts: (A1; partially) (C1) (C3)
Increasing volume of transportation: (A1; ballast) (B4) (C5; aerial deposition)
Leakage from industrial processes: (B1)
Multiple-source stress: (C1) (C4) (C5)
Uncontrolled societal development: (C7)

Mitigation
Innovation in research methods: (B1) (C4) (C5) (C6)
Technological innovation: (B7)
Informing the public: (B2; recommendations on eating fish)
Specific management practices: (A2; removal) (B2) (B7) (B8)
Risk management: (B1) (B4)
Administrative regulation and control: (A1) (B1) (B5) (C1) (C3)

Note: See this chapter for an explanation of the codes.
Source: Leppäkoski et al (2006); Karjalainen et al (2006); Kuosa et al (2006)

The next step in investigating the problem space is to construct a representation of it by using the categorizations in Table 9.1. An ordinary Cartesian representation is out of the question for two main reasons: first, the dimensions are not independent of each other – far from it. Second, it would be extremely awkward to represent the categories as some sort of a continuum along orthogonal axes. However, what can be reasonably done is to compare the position of the problem categories along the dimensions, from more specific to more general (see Figure 9.1).

Figure 9.1 indicates that there is, indeed, some structural order in the relative positions of the problem categories along the three axes. Several problems are located consistently to the left along the two first axes, and also to the left of the corresponding mitigation measure on the third axis (e.g. 'administrative control' covers 'purposeful human acts'). These problems might then be candidates for achieving closure. On the other hand, on the right-hand side there is critical variation: there are no mitigation measures that could adequately respond to the most general 'target' and 'source' categories. It would seem that in the case of these problems, closure is a distant dream.

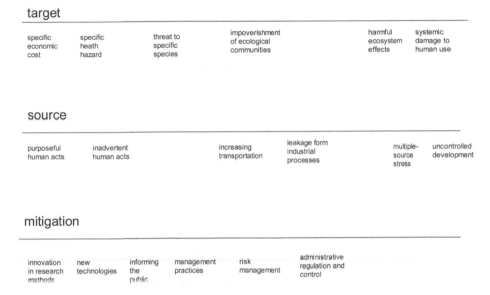

Figure 9.1 *The relative locations of problem categories (see Table 9.1) along the axes characterizing the eco-social problem space related to the Baltic Sea, according to the source articles (Leppäkoski et al, 2006; Karjalainen et al, 2006; Kuosa et al, 2006)*

Note: The categories are ordered from more specific to more general (left to right).

Source: Leppäkoski et al (2006); Karjalainen et al (2006); Kuosa et al (2006)

HETEROGENEITIES

The exercise thus far has a critical shortcoming: the representation in Figure 9.1 looks uniform and static; but we know that the categories are heterogeneous and dynamic.[2] Two important types of heterogeneity can be identified in the source articles in *Vesitalous* (vol 47, no 2, 2006): spatial scales and variation in types of knowledge. A third type shows up in the case studies presented in Chapter 8 and Peuhkuri (2002, 2004): variation in social and political context. These are taken up one at a time.

Eco-social scales

The need for appropriate scaling is straightforward. Meadowcroft (2002, p169) makes the case succinctly (see also Haila, 2002; Haila and Dyke, 2006):

> *'Limit' and 'scale' are, after all, closely interconnected. A limit is a boundary beyond which certain forces, processes or rules no longer apply. In other words, it is a boundary beyond which scale matters: for*

what held true within the limit can no longer be relied upon to hold
true once the threshold is crossed.

Put this way, the issue is so obvious that it is quite natural that questions of scaling prop up in the source articles in *Vesitalous* (vol 47, no 2, 2006) in several forms (see Table 9.2).

Table 9.2 *Aspects of eco-social scaling taken up in the source articles (Leppäkoski et al, 2006; Karjalainen et al, 2006; Kuosa et al, 2006)*

(A1) Differentiation of conditions across the Baltic basin

Variation in coastal geomorphology and ecology
Salinity of the water drops from around 20 parts per thousand to basically freshwater in coastal bays in the north
The discharge from different river basins varies substantially depending upon geographical conditions
There is variation in the dynamics of internal load and susceptibility to eutrophication (see Bonsdorf et al, 2002; Rönnberg and Bonsdorf, 2004)

(A2) Scale-crossing events

Increasing shipping → more introduced species
Increasing volume of oil transportation → increasing risk of tanker accidents
Increasing car traffic → increasing load of airborne nitrogen

(A3) Temporal change in the nature of the problems

The problems change shape in time, to a large extent as a function of success or failure of environmental management
Successful nature conservation changes the agenda of conservation (see Chapter 8 in this volume)

Source: Leppäkoski et al (2006); Karjalainen et al (2006); Kuosa et al (2006)

Types of knowledge

Knowledge is taken up in several different guises in the source articles: the authors refer both to focused experimental results (removal of mink) and to broad-scaled systems models (ecosystem model of the Baltic). To get a systematic overview of what kind of knowledge is used as arguments in the papers, this section identifies three different aspects of knowledge following Haila and Dyke (2006); this owes much to Michel Foucault's work (Foucault, 1972, 1980).

Foucault (1972) draws a useful distinction between two dimensions of knowledge that are expressed by different terms in French – namely, *savoir* and *connaissance*. The first term refers to matters of fact (B1) as something specific that can be known; the second term refers to systems of knowledge (B2) that give meaning to particular matters of fact. With regard to scientific knowledge, the distinction is obvious enough: we know 'facts' and we have 'theoretical frameworks' that give scientific meaning to these facts.

A second specification adds the social dimension to the picture. Scientific facts and theories do not spell out their relevance for human affairs all by themselves. Quite the contrary, assessing the relevance of any 'matter of fact' or 'theoretical frame' is a social interpretation and requires a normative framework (B3). Environmental concerns offer pertinent examples: they arise because some matters of fact about human-induced changes in the environment are interpreted as

Table 9.3 *References to different types of knowledge in the source articles (Leppäkoski et al, 2006; Karjalainen et al, 2006; Kuosa et al, 2006)*

(B1) Matters of fact

Removal of mink → improved nesting success of water birds and shorebirds
Grazing of 'opportunistic' algae → stabilization of algal community
Concentrations of hazardous chemicals in organisms of the Baltic Sea
Toxicity of dioxin (to humans)
Effects of toxic oil residues on organisms
Biomanipulation → improvement in the state of eutrophic lakes
Environmental chemistry of phosphorus and nitrogen in the Baltic Sea
Mechanisms of internal loads, including interaction with external loads

(B2) Systems of knowledge as interpretative frameworks

Biogeographic history of the Baltic Sea
Predator–prey interactions (mink versus waterfowl)
Mosaic dynamics in the algal community of the littoral zone
Transparency of water as a 'forcing function'
The knowledge base of risk management
Ecosystem model of the Baltic Sea

(B3) Normative frameworks

Framework for normative assessment: natural versus modified:

> *The BIREME research project has investigated changes in the biotic species of the Baltic Sea over shorter and longer timelines, from the early arrival of species to recent changes. (Leppäkoski et al, 2006, Abstract)*

Framework for normative assessment: toxic compounds present versus those that are absent:

> *Due to anthropogenic activity, foreign compounds, some highly toxic, have entered the environment in large amounts in recent years.(Karjalainen et al, 2006, Abstract)*

Framework for normative assessment: the Baltic Sea is ecologically healthy versus environmentally degraded:

> *The greatest problem continues to be eutrophication. Efforts to solve this problem have forced us to recognize deficiencies in our knowledge of the factors that affect the wellbeing of the Baltic Sea. (Kuosa et al, 2006, Abstract)*

Source: Leppäkoski et al (2006); Karjalainen et al (2006); Kuosa et al (2006)

problems. The assessment of the social relevance of matters of fact is normative. For instance, environmental concerns stipulate that all human-induced changes in the environment need assessment.[3]

I turned to the articles in *Vesitalous* (vol 47, no 2, 2006) to identify different types of knowledge claims made by the authors (for a summary of the results, see Table 9.3). A clarifying note is required. The two first types – (B1) and (B2) – can, strictly speaking, be defined in relation to ongoing research practices that maintain and reproduce them constantly. Hacking (1992) refers to this process as 'stabilization of laboratory research'; this notion is also applicable to ecological field studies (Haila, 1998). The third type (B3), by contrast, is not directly tied to any research practice. It reflects general, perhaps unarticulated, beliefs about the relevance of the type of research in question, shared by the scientists who do the work but also by a broader circle of people who are interested in the results.

Social and political constituencies

The fact that environmental problems are constructed (i.e. that they must be *defined* and *recognized* as problems by social actors) has two major implications.

First, although science (the agency of scientists) is usually a necessary condition for environmental problems to be defined as problems at all (e.g. Hannigan, 1995), this is not sufficient. The problem definition needs societal support: a constituency (i.e. it needs to become public). I have earlier used the metaphor that issues carve their social space through the comprehensiveness and depth of mobilization that they force upon human actors within their sphere (Haila, 2002, 2004). This process has, of course, been going on in varying forms ever since the environmental awakening of the 1960s.

Second, every newly defined problem enters into competition with older problem definitions. There is no 'clean table' from where to begin. Furthermore, in this competition between different problem definitions, the problems do not face each other in their sheer materiality. On the contrary, the critical factor is their relative symbolic weight. The success of a particular problem definition in winning a broad constituency depends upon how efficiently it resonates with the aspirations and desires of those who might potentially join in.

Pierre Bourdieu's criticism of objectivism that I took up above is directed against the neglect of the symbolic dimension of the social world. The context of the remark is worth citing in full (Bourdieu, 1991, p229):

> *The construction of a theory of the social space presupposes ... a break, finally, with objectivism, which goes hand in hand with intellectualism, and which leads one to overlook the symbolic struggles that take place in different fields, and where what is at stake is the very representation of the social world, and in particular the hierarchy within each of the fields and between the different fields.*

Bourdieu's notion of 'field' refers to a space of social action in which people interact with each other and compete for power and prestige; it is 'an autonomous universe, a kind of arena in which people play a game which has certain rules, rules which are different from those of the game that is played in the adjacent space' (Bourdieu, 1991, p215). Using this terminology we can say, for instance, that once a shared perception of environmental problems is established in society, an 'environmental field' gradually emerges. Since then there has been competition (or struggle) about the 'correct' definition of environmental problems within this field, and there has been competition (or struggle) about the position of the environmental field with respect to other, older, well-established fields.

Now back to Bourdieu's view of the importance of symbolic struggles: that 'what is at stake is the very representation of the social world'. From the perspective of environmental concerns, the statement can be put like this: what is at stake is the relevance of environmental problems in the representation of the social world.

With regard to environmental problems in the Baltic Sea Region, the next question to ask is: where is the environmental field? Who are the 'players' in the field? What, precisely, are the issues that are of concern?

At this point, we have to leave *Vestralous* (vol 47, no 2, 2006) behind and turn to other sources. On a general level, the threatening state of the Baltic Sea is today a major concern among the general public in all countries that border it (see Chapter 4 in this volume). This is due to the considerable publicity that the issue has received, but perhaps even more importantly to the broad range of concrete inconveniences and practical troubles that eutrophication of the sea has caused to ordinary people (this point is included in Table 9.1).

It is, however, also certain that the environmental field becomes more complex when we get closer to the heterogeneous set of issues listed in Table 9.1. To get a glimpse of the differentiation involved, we can turn to Chapter 8 in this volume and Peuhkuri (2002, 2004).

The topic of Chapter 8 is the conflict between seal conservation and coastal fishery in the Kvarken region. The conflict is fairly strictly localized and concerns coastal communities in two countries – Finland and Sweden – on both sides of the Kvarken. The conflict originated as a result of successful seal conservation. The search for a solution that Chapter 8 analyses became possible when a new coalition of actors was formed on the regional level, partially side stepping established authorities on both national and local levels. As a consequence, new governance arrangements have gradually emerged, and a new constituency has been formed. The dynamic reframing of the issue has also given rise to new problem definitions – for instance, the challenge of developing a seal-proof trap net.

The topic of Peuhkuri's (2002, 2004) research project is a long-term conflict concerning the effect of fish farming on the eutrophication of the Archipelago Sea. The situation has received much publicity in the national media. Peuhkuri has focused, in particular, on the role of knowledge in the discursive interpretation of the conflict. He found out that all main stakeholders – local inhabitants and fish farmers, local politicians, regional and national environmental authorities, and environmental scientists (mainly ecologists) – largely agreed on the importance of the problem; but assessing its severity raised controversies in concrete contexts. In a nutshell, the question was: how large an influence does fish faming, a local source of eutrophication, have on the scale of the Baltic as a whole?

Peuhkuri was able to identify three competing framings of the issue: 'living archipelago', 'endangered sea' and 'technical management'. The first framing emphasized local cultural values and livelihood, whereas the two others took as their starting point a trans-local science-based assessment of the severity of the situation as a whole. This difference in perspectives produced a binary setting, which reduced the chances of reaching a mutual agreement on the nature of the problem. The conflict is still largely open, although some authorities are developing an understanding of local livelihood needs.

Local and trans-local problem framings are in close interaction in the traject-ory of the conflicts analysed by Chapter 8 and by Peuhkuri (2002, 2004). General environmental concerns provide backing for sentiments that do not take into account local livelihood needs ('Don't interfere with the seals!'; 'Stop fish farming in the endangered Archipelago Sea!'). Local people, by contrast, view themselves as inheritors of a 'long tradition of living from nature' (Peuhkuri, 2002, p167). Through this framing, an element of justice is introduced to the conflict: local people demand respect for their livelihood needs, including the right to modernize their subsistence practices. This sort of conflict remains unsolved as long as the perspective of either party is not recognized as legitimate.

Problem closure? An assessment

There is enormous heterogeneity in factors that define different types of envir-onmental problems all over the Baltic Sea Region. This conclusion in itself is not particularly surprising or new. However, for the development of environ-mental governance it is useful, if not necessary, to get a structured view of this heterogeneity.

Let's start from problem closure. What can be said about the similarity of criteria that are used in assessing problems along the three dimensions? To what extent can the problems and potential solutions be defined with the same criteria (using the operational definition of closure mentioned in the introduction to this chapter)?

In the case of problems located on the left-hand side of Figure 9.1, closure seems possible. However, closure is not an automatic result of problem definition, not even for problems that are well focused. Closure is the end point of a circular movement, which converges towards stable criteria. With reference to Figure 9.1, we could envisage this circular movement as a process of entrainment that captures criteria defined along the three separate axes, and the result is a coherent way of defining the problem (for 'entrainment', see Dyke, 2006). 'Shared cognition' (Hutchins, 1995) would be a useful image for a stabilized problem closure that presupposes coordinated action by several actors. The focus shifts continuously in a thoroughly interactive and non-linear way (Dyke, 1994). Since there is a strong element of practical management involved in this process, the idiom of 'co-production of knowledge and governance' is illuminating. As Sheila Jasanoff (2004, p2) puts it: 'co-production is a shorthand for the proposition that the ways in which we know and represent the world (both nature and society) are inseparable from the ways in which we choose to live in it'.

For instance, the problem of oil spills is amenable to closure nowadays; but this was not the case during the 1970s. This change has required quite a lot of technical development, new regulatory and control practices and institutional arrangements (such as facilities for discharging waste oil at harbours). Scale-crossing effects are still a problem in the case of oil spills, however. There is not much that environmental authorities within the Baltic Sea Region can do about the international trade in crude oil. On the other hand, some relatively simple problems lack closure – for instance, established alien species such as the barnacle: there is no way of getting rid of this nuisance; we just have to carry the costs.

As to the diffuse problems located to the right of Figure 9.1, adequate problem closure is not within sight at the moment. It seems advisable to cut the problems into more manageable pieces, and address them one at a time (see also Bonsdorf et al, 2002). Basically, this amounts to a process of framing and reframing the problems in different ways and from different perspectives so that adequate problem-specific constituencies will take shape.

Sensitivity to the spatial context and to the social and political situation helps a lot. This claim may seem counter-intuitive to specialists in the natural sciences: doesn't it create more complications if we broaden the range of aspects that need consideration and include socio-economic issues? In fact, we can draw the opposite conclusion: specific problems have specific constituencies, and problem closure is achievable only if the constituencies get involved in the work. Solutions of specific environmental problems require practical efforts and lots of local participants. Just think of biomanipulation of eutrophic lakes upstream. The task is impossible on any relevant scale without the participation of a broad range of local actors.

The case of seal conservation in the Kvarken region shows that an adequate constituency often has to be created on the spot; it does not exist in advance. The creation of the constituency may involve controversies, and natural science is only one factor to be taken into account. This is the basic take-home message from Peuhkuri's (2002, 2004) case study of fish farming and eutrophication in the Archipelago Sea. The actual threat caused by a specific activity such as fish farming has to be assessed in an adequate spatial and temporal scale, and the measures to be taken have to be planned accordingly. There is no choice in this respect. Such a process of assessment has started, quite successfully, in the conflict between seal conservation and coastal fishery in Kvarken.

As environmental problems are woven together with the values and livelihood of local people, they are only exceptionally amenable to an 'optimal solution' that could be derived from science alone. It is more important to get the right people involved, and to respect the identities and subsistence needs of local people than to make calculations on some abstract grounds. This amounts to a continuous process of framing and reframing, with the deliberate purpose of getting the relevant stakeholders recruited into the constituency.

By way of a conclusion: to reach problem closure, we need to stabilize the field. To stabilize the field, we need to get the right constituency involved in practical work. This usually requires reframing the issue in adequate ways (Hajer, 2003b). Finally, it is important not to pretend that closure can be easily achieved in a multiple-stress situation, such as the deterioration of the Baltic Sea Region.[4]

GOVERNANCE: GENERALITY VERSUS SPECIFICITY

As noted in the introduction to this chapter, strong factors have driven towards the unification of environmental problems. Originally, the need for unification was a rhetorical strategy, adopted, by necessity, by people advocating the environmentalist cause in public protest movements and in governmental circles. The drive towards unity was also motivated by the need to gain symbolic weight for environmental concerns, in general. The older obligation of nature conservation joined forces with environmental concerns, and what we have today are two broad normative frameworks that provide apparently unified perspectives to environmental issues: sustainable development and biodiversity.

However, the reality of the problems presents us with remarkable heterogeneity: how to deal with the stark contrast between the heterogeneity of the problems and the perceived need of unification of the policy response?

This question has certainly been addressed in environmental policy literature. One useful strand has been to elaborate upon general conditions of making environmental governance plausible and legitimate in the first place. For instance, Lafferty and Meadowcroft (1996) evaluate the importance of democracy as a general precondition of environmental policy, and Kenny and Meadowcroft (1999) and Lafferty and Meadowcroft (2001) analyse the requirements of implementing principles of sustainability in public policy.

Another useful resource is offered by policy analysis focusing on the political process, particularly policy practices that are (or are not) conducive to successful implementation of progressive policies of any kind. The aim is to 'shift the debate about democracy from the normative to the empirical' (Hajer and Wagenaar, 2003, p3). This approach emphasizes definitional struggles in the understanding of political process; implementation of specific policies on the ground is understood as a critical element in the process. The idea of problem closure can be integrated within this approach as a semi-stable end result of problem framing (Hajer, 2003b; Laws and Rein, 2003).

In other words, important elements needed for dealing with the generality versus specificity issue are at hand: the general structure of environmental governance is a stable element, in the background, as defined, for instance, by EU legislation and directives. This gives certain space for the development of multilevel structures and the involvement of different constituencies, as discussed in Chapters 1 and 10 in this volume.

However, it seems to me that further specification is in order. We can approach a specific but complex task, such as protection of the Baltic, from two sides. The approach from above takes as its framework an analysis of the whole system: the task is to identify the worst threats and tackle them (e.g. the nutrient load from St Petersburg). The approach from below, on the other hand, takes as its framework an analysis of specific situations: the task is to differentiate management practices and get the relevant constituencies involved. Of course, the problem of wastewater treatment in St Petersburg has this bottom-up aspect, too, in the task of getting an adequate constituency to take shape (see Tynkkynen, forthcoming).

My basic emphasis would be on bottom-up processes. On the ground level, the main challenge is to make alternatives visible, as well as to create new alternatives (Haila, 1999). On this level, eco-social problem space joins hands with eco-social possibility space: the task is to broaden the possibility space (for more on this notion, see Dyke, 1988). There are choices to be made between alternatives, which have to be specified – in fact, discovered, as in the Kvarken case (see Chapter 8).

The emphasis of bottom-up perspectives implies a reassessment of the role of knowledge in environmental issues. There is a strong emphasis on the need for more research in the source articles in *Vesitalous* (vol 47, no 2, 2006). In some sense, this is natural (in a Bourdieuean mode we might note that this is a necessary aspect of the activity of scientists within their own field). However, if taken strictly literally, the demand for ever more research may be a recipe for disaster. Western science could well be the most conspicuous example of listening to some things at the expense of desensitizing ourselves to other, more important, things (Haila and Dyke, 2006, p38). Science is invaluable; but scientific distancing and objectification has to be complemented with tacit knowledge about socio-cultural settings.

Perhaps we know enough about many of the problems. Perhaps the main challenge is to broaden the conception of knowledge to include the practical wisdom of local actors whose livelihood is at stake. Solutions imposed from above cannot but induce mistrust, resistance and conflicts. The call for more scientific research is a reflection of the rationalistic dream of total control. Concerning a target as complex as the eco-social system of the Baltic Sea Region, 'total control' is a pipe dream, whether on the basis of knowledge or on some other grounds. Instead of dreaming of a rationalistic control of this system, we need, paraphrasing Foucault (2000, p329), to analyse the specific rationality of the environmental field and to develop problem-specific approaches that respect existing experience and knowledge.

NOTES

1   The profile of this journal is indicated by the subtitle on the page where English summaries of the articles are printed: 'Finnish journal for professionals in the water sector'.
2   Uniformity is, of course, a characteristic of a Cartesian phase space representation: the axes seem static while, in fact, they stand for dimensions that are dynamic. We could get rid of this misperception by using more complicated topological shapes; but that would stretch the analogy too far.
3   Bruno Latour (2004) draws a similar distinction between what he calls 'matters of fact' and 'matters of concern'.
4   A comparison with the Kyoto-driven international climate policy is interesting on this point. If we accept the reduction of climate policy to carbon dioxide policy, the protection of the Baltic Sea appears as a more complex governance challenge than climate change. Whether this reduction is, ultimately, productive is another matter.

REFERENCES

Bonsdorff, E., Rönnberg, C. and Aarnio, K. (2002) 'Some ecological properties in relation to eutrophication in the Baltic Sea', *Hydrobiologia*, vol 475/476, pp371–37.
Bourdieu, P. (1991) *Language and Symbolic Power*, Polity Press, Oxford
Dyke, C. (1988) *The Evolutionary Dynamics of Complex Systems: A Study in Biosocial Complexity*, Oxford University Press, New York
Dyke, C. (1993) 'Extralogical excavations: Philosophy in the age of shovelry', in J. Caputo and M. Yount (eds) *Foucault and the Critique of Institutions*, Pennsylvania State University Press, Philadelphia, pp102–126
Dyke, C. (1994) 'The world around us and how we make it: Human ecology as human artefact', *Advances in Human Ecology*, vol 3, pp1–22
Dyke, C. (2006) 'Primer: On thinking dynamically about the human ecological condition', in Y. Haila and C. Dyke (eds) *How Nature Speaks: The Dynamics of the Human Ecological Condition*, Duke University Press, Durham, NC, pp279–301
Foucault, M. (1972) *The Archaeology of Knowledge*, Pantheon, New York
Foucault, M. (1980) *Power/Knowledge: Selected Interviews and Other Writings 1972–1977*, Pantheon, New York
Foucault, M. (2000) 'The subject and power', in J. D. Faubion (ed) *Essential Works of Foucault 1954–1984, vol 3: Power*, The New Press, New York, pp326–348
Garfinkel, A. (1981) *Forms of Explanation: Rethinking the Questions in Social Theory*, Yale University Press, New Haven
Gren, I.-M., Wulff, F. and Turner, K. (eds) (2000) *Managing a Sea: The Ecological Economics of the Baltic*, Earthscan, London
Hacking, I. (1992) 'The self-vindication of the laboratory sciences', in A. Pickering (ed) *Science as Practice and Culture*, University of Chicago Press, Chicago, pp29–64
Hacking, I. (2000) *The Social Construction of What?* Harvard University Press, Cambridge, MA

Haila, Y. (1998) 'On the political undercurrents of modern ecology', *Science as Culture*, vol 7, pp465–491

Haila, Y. (1999) 'Socioecologies', *Ecography*, vol 22, pp337–348

Haila, Y. (2002) 'Scaling environmental issues: Problems and paradoxes', *Landscape and Urban Planning*, vol 61, pp59–70

Haila, Y. (2004) 'Making sense of the biodiversity crisis: A process perspective', in M. Oksanen and J. Pietarinen (eds) *Philosophy of Biodiversity*, Cambridge University Press, Cambridge, pp54–83

Haila, Y. and Dyke, C. (2006) 'Introduction: What to say about nature's "speech"', in Y. Haila and C. Dyke (eds) *How Nature Speaks: The Dynamics of the Human Ecological Condition*, Duke University Press, Durham, NC, pp1–48

Haila, Y. and Levins, R. (1992) *Humanity and Nature: Ecology, Science and Society*, Pluto Press, London

Hajer, M. (2003a) 'Policy without a polity: Policy analysis and the institutional void', *Policy Sciences*, vol 36, pp175–195

Hajer, M. (2003b) 'A frame in the fields: policy-making and the reinvention of politics', in M. A. Hajer and H. Wagenaar (eds) *Deliberative Policy Analysis: Understanding Governance in Network Society*, Cambridge University Press, Cambridge, pp88–110

Hajer, M. and Wagenaar, H. (2003) 'Introduction', in M. A. Hajer and H. Wagenaar (eds) *Deliberative Policy Analysis: Understanding Governance in Network Society*, Cambridge University Press, Cambridge, pp1–30

Hannigan, J. A. (1995) *Environmental Sociology: A Social Constructionist Perspective*, Routledge, London

Hutchins, E. (1996) *Cognition in the Wild*, MIT Press, Cambridge, MA

Jansson, B.-O. and Hammer, M. (1999) 'Patches and pulses as fundamental characteristics for matching ecological and cultural diversity: The Baltic Sea archipelago', *Biodiversity and Conservation*, vol 8, pp71–84

Jansson, B.-O. and Velner, H. (1995) 'The Baltic: The sea of surprises', in L. Gunderson, C. S. Holling and S. S. Light (eds) *Barriers and Bridges to the Renewal of Ecosystems and Institutions*, Columbia University Press, New York, pp292–372

Jasanoff, S. (2004) 'The idiom of co-production', in S. Jasanoff (ed) *States of Knowledge: The Co-production of Science and Social Order*, Routledge, London, pp1–12

Karjalainen, J. et al (2006) 'Threats posed to the Baltic Sea by toxins – two examples' (in Finnish with an abstract in English), *Vesitalous*, vol 47, no 2, pp15–19

Kenny, M. and Meadowcroft, J. (eds) (1999) *Planning Sustainability*, Routledge, London.

Kuosa, H. et al (2006) 'Variable nutrient retention in the catchment area and complex nutrient dynamics in the Gulf of Finland complicates the recovery of the Baltic Sea ecosystem' (in Finnish with an abstract in English), *Vesitalous*, vol 47, no 2, pp20–25

Lafferty, W. and Meadowcroft, J. (eds) (1996) *Democracy and the Environment: Problems and Prospects*, Edward Elgar, London

Lafferty, W. and Meadowcroft, J. (eds) (2001) *Implementing Sustainable Development: Strategies and Initiatives in High Consumption Societies*, Oxford University Press, New York

Latour, B. (2004) 'Why has critique run out of steam? From matters of fact to matters of concern', *Critical Inquiry*, vol 30, winter, pp225–248

Laws, D. and Rein, M. (2003) 'Reframing practice', in M. A. Hajer and H. Wagenaar (eds) *Deliberative Policy Analysis: Understanding Governance in Network Society*, Cambridge University Press, Cambridge, pp172–206

Leppäkoski, E., Helminen, H., Hänninen, J. and Tallqvist, M. (1999) 'Aquatic biodiversity under anthropogenic stress: An insight from the Archipelago Sea (SW Finland)', *Biodiversity and Conservation*, vol 8, pp55–70

Leppäkoski, E., Väinölä, R., Korpimäki, E. and Jormalainen, V. (2006) 'Biological diversity of the Baltic Sea – only change is permanent' (in Finnish with an abstract in English), *Vesitalous*, vol 47, no 2, pp10–14

Meadowcroft, J. (2002) 'Politics and scale: Some implications for environmental governance', *Landscape and Urban Planning*, vol 61, pp169–179

Peuhkuri, T. (2002) 'Knowledge and interpretation in environmental conflict: Fish farming and eutrophication in the Archipelago Sea, SW Finland', *Landscape and Urban Planning*, vol 61, pp157–168

Peuhkuri, T. (2004) 'Roles of knowledge in environmental conflict: Debate and decision making concerning eutrophication and fish farming industry in the Archipelago Sea Region, in southwest Finland' (in Finnish with an abstract in English), *Annales Universitatis Turkuensis*, Ser. C, no 220

Rapport, D. J. (1989) 'Symptoms of pathology in the Gulf of Bothnia (Baltic Sea): Ecosystem response to stress from human activity', *Biological Journal of the Linnean Society*, vol 37, pp33–49

Rönnberg, C. and Bonsdorff, E. (2004) 'Baltic Sea eutrophication: Area-specific ecological consequences', *Hydrobiologia*, vol 514, pp227–241

Tynkkynen, N. (forthcoming) 'Environmental cooperation and learning: The St Petersburg water sector', in D. Lehrer and A. Korhonen (eds) *Western Aid in Postcommunism: Effects and Side-effects*, Palgrave Macmillan, Basingstoke

Wulff, F., Bonsdorff, E., Gren, I.-M., Johansson, S. and Stigebrandt, A. (2001) 'Giving advice on cost effective measures for a cleaner Baltic Sea: a challenge to science', *Ambio*, vol 30, pp254–259

Part V

Conclusions

10

Governing a Common Sea: Comparative Patterns for Sustainable Development

Kristine Kern, Marko Joas and Detlef Jahn

INTRODUCTION

The previous chapters have shown that new forms of governance have been developed and tested in the Baltic Sea Region. We can now ask whether the Baltic Sea Region can serve as a model for other comparable regions, both in Europe and throughout the world. The Baltic Sea shares certain characteristics with other regional seas, such as the Mediterranean Sea and the Black Sea. All three regional seas are highly vulnerable to pollution and are surrounded by states with differing economic, political and cultural context factors (for the Baltic Sea Region, see Chapters 2, 4 and 5 in this volume). Furthermore, all three regions have been directly affected by European integration. Apart from the implications of our research results for other regional seas in Europe, the lessons learned in the Baltic Sea Region could also be relevant for other regional seas around the world, from the Caspian Sea to the Great Lakes in North America and the Great Lakes in Africa.

In this concluding chapter we first discuss and summarize the various forms of governance for sustainable development in the Baltic Sea Region. The Baltic Sea Region is compared to other regional sea areas in the following section. This comparison serves as a basis for highlighting institutional arrangements in the Baltic Sea Region that are unique but may be transferred to other regional sea areas where they could help in resolving persistent problems.

GOVERNANCE FOR SUSTAINABLE DEVELOPMENT IN
THE BALTIC SEA REGION

Both traditional and new forms of environmental governance can be found in the Baltic Sea Region. More traditional forms are observed at the national level. National environmental governance in the riparian states varies considerably because the region comprises the Nordic countries, reunited Germany and former socialist countries, which faced enormous challenges after the fall of the Berlin Wall. When focusing on the governance of the Baltic Sea as a common good, three forms of governance appear to be of special interest: first, traditional forms of international environmental governance, such as the Helsinki Convention on the Protection of the Marine Environment of the Baltic Sea Area; second, new forms of European governance have become very important for the sustainable development of the region following the EU enlargements of 1995 (Sweden and Finland) and 2004 (Poland, Lithuania, Latvia and Estonia); and, finally, various forms of transnationalization can be observed in the Baltic Sea Region. The establishment of Baltic 21 appears to be most important for sustainable development because this Agenda 21 process for the Baltic Sea Region involves not only nation states but also transnational actors and includes transnational networks of both sub-national governments and civil society actors which have emerged and developed dynamically in the region since the end of the Cold War.

Basic pollution and policy-setting in the Baltic Sea Region

The state of the environment and the environmental condition of the Baltic Sea, in particular, pose a crucial challenge to the political systems surrounding the Baltic Sea. The severe political and economic division of the region is clearly visible in the relationship between economic performance and environmental impacts. The development of economic growth and environmental pollution in the Eastern riparian countries does not follow the trend of the Western riparian countries in the Baltic Sea Region. The decoupling of economic growth and environmental pollution occurs at a much earlier stage of economic development in the Eastern riparian countries than in Western riparian countries (see Chapter 2 in this volume).

However, there are differences with respect to how and when environmental issues reached the political agenda and how the systems responded to and acted on environmental challenges. In Chapter 4, Hermanson notes that the shift in values towards more post-materialist attitudes that has occurred in advanced industrial societies is also visible in the Baltic Sea Region, but only to a certain extent. In the Nordic countries and (West) Germany, social movements began to put environmental issues on the political agenda as early as the 1970s. They influenced decisions initially at the local level, became more organized and eventually challenged political decision-making at the national level. Both this organizational transformation and the 'environmental awakening', which triggered public awareness of new types of environmental problems, started at around the same time. A new perception of the society–environment relationship developed due to social changes (see Chapter 9 in this volume) and, as a result, environmental policy was established as a new sector of public policy.

Due to the differences in the political systems, this political institutionalization of environmental concerns occurred much later in the Baltic states, Poland and Russia. Although environmental movements played an important role in some Central and Eastern European (CEE) countries when the Soviet empire finally collapsed, the division between the two groups of countries in relation to environmental/post-material values is still apparent.

National environmental governance

Having established governmental institutions for environmental protection and national strategies at an early stage, the Nordic countries and (West) Germany have been clear forerunners in the process of institutionalizing environmental concerns (see, for example, Jänicke and Weidner, 1997; Hermanson and Jahn, 1998). The conditions in countries experiencing rapid transition, such as Russia, Lithuania, Estonia and Latvia, have been quite different. The development of modern environmental policy in these countries started from a centrally planned economy, with institutions suited for this purpose, state-owned property, a one-party system and limited public debate on social and political issues.

In the Baltic states, transition influences both environmental policy-making and society as a whole (according to Chapter 5 in this volume). The time frame for changes is different compared with Western European and Nordic countries, where environmental policy and governance, on the one hand, and environmental awareness, on the other, developed gradually and concurrently over several decades. One feature is, however, striking. The shift towards modern European Union-style 'standard' government in environmental policy has been rapid and, naturally, guided by the EU enlargement process.

According to Meadowcroft (2007, pp161–162), several European countries, among them many if not most countries in the Baltic Sea Region, are taking sustainability planning seriously and are thus moving a step further away from environmental policy-making towards integrated policy-making for sustainable development. This shift also includes a change in the way that government processes work. The empirical change towards new modes of governance which involve a higher number of actors and more stakeholder groups in decision-making is clearly visible in the whole region. This change is not only apparent at the international level; it is also evident in national policy-making and is, in fact, becoming more and more evident at the local level.

Steuer (2007, p208) characterizes sustainable development as one of the most obvious changes in the patterns of general national governance from bureaucratic planning to new public management and a market-based era, and, finally, to a society developing new governance modes, based on the primary notion of calibrating institutions and societal processes to work for common goals. The new governance institutions favour an inter-organizational focus over an intra-organizational one. These new forms of environmental governance involve an increasing policy-making role for municipalities and other local government bodies, as well as various non-governmental organizations and networks.

At the local level, many new modes of governance for sustainable development were directly related to the development of Local Agenda 21 (LA21) processes. The emergence of LA21 can be regarded as a success story from a Baltic Sea perspective because the new structures diffused fairly rapidly both within countries and across former geopolitical borders (see Chapter 7 in this volume). While the Nordic countries, and particularly Sweden, reacted shortly after the United Nations Conference on Environment and Development in 1992 and soon became international pioneers, most LA21 processes in reunited Germany were initiated much later and spread quite differently throughout the German federal states. Although cities located in former East Germany lagged behind the rest of the country until recently, the clear East–West divide has disappeared. Most LA21 processes that were launched only recently have been initiated in eastern states such as Mecklenburg-Western Pommerania (Kern et al, 2007). LA21 processes in Poland and in the Baltic republics share some similarities with their East German counterparts, although the development of LA21 in East Germany appears to be more dynamic.

Interpreting regulations and educating actors has become an important feature of network governance because environmental regulations tend to be more detailed and target a higher number of diverse actors and individuals. Today's general nature conservation goals are often formulated in international laws and conventions. From an idealistic point of view, at least, it is assumed that international-level decisions are ratified by national parliaments and the ratifying states then organize the actual implementation (from planning to enforcement) of conservation policies in a hierarchical way. This idealistic perspective can cause serious problems because hierarchical governance has its limits in many policy areas. Difficult situations can occur if conservation goals can only be fulfilled by restricting existing and well-established activities, while the citizens whose activities are influenced do not share the conservation goals or the perception that their activities may influence the pursuit of these goals. Incompatible interests in environmental resources may result in environmental conflict. Mediation, often through the development of networks, will help in attaining a higher level of legitimacy and understanding for environmental regulation and create innovations designed to cope with the governance of complex multilevel issues (see Chapter 8 in this volume).

International and intergovernmental cooperation

As a matter of fact, transnational and international connections have always existed in the Baltic Sea Region. These links can be traced back to the Hanseatic League and, surprisingly, even survived the Cold War period. Environmental concern about increasing pollution of the Baltic Sea was among the driving forces of international environmental cooperation in the Baltic Sea Region, although political factors played an essential and decisive role here (see Chapter 3 in this volume). While the Nordic countries always had close links with each other, which have been institutionalized in the Nordic Council, international environmental cooperation between all of the riparian countries of the Baltic Sea rendered the Iron Curtain somewhat transparent.

Most important in this context has been the Helsinki Convention with the Helsinki Commission as its governing body (HELCOM). The goal of HELCOM is to protect the marine environment of the Baltic Sea from all sources of pollution, and to restore and safeguard its ecological balance. In 1974, the then seven Baltic coastal states signed a convention for the prevention and abatement of all sources of pollution around the Baltic Sea. It was ratified by all seven countries and entered into force in 1980. After the collapse of the Soviet Union, the reunification of Germany and the attainment of independence by the Baltic states, all states bordering the Baltic Sea and the European Community agreed on a new convention in 1992, which entered into force in January 2000. The area of concern was extended and now includes inland waters, the seawater and the sea bed. Thus, measures are taken in the entire catchment area of the Baltic Sea to reduce land-based pollution.

Soon after the EU enlargement of 1995, when Finland, Sweden and Austria joined the European Union, the Finnish government initiated the 'Nordic Dimension' as a regional EU strategy that established a partnership between the EU, Norway, Iceland and Russia. The revision of the Northern Dimension was launched in 2006, transforming it into a common regional policy. The new Northern Dimension has four geographical priorities: the Baltic Sea, Kaliningrad, the Barents Seas and the Arctic. It will increasingly focus its activities on northwest Russia.

The Nordic Dimension is broad in its scope, but places strong emphasis on environmental issues. The Northern Dimension Environmental Partnership (NDEP), which was established in 2001 by various international financial institutions, such as the European Bank for Reconstruction and Development (EBRD), the Nordic Investment Bank (NIB) and the World Bank, reflects this initiative. One major concern of the NDEP is improving the environment and managing spent nuclear fuel in Russia, which is currently a weakness in environmental protection in the Baltic Sea Region. In this context, the NDEP focuses, in particular, on Kaliningrad, the Russian enclave now surrounded by EU member states.

European Union integration and Europeanization

EU integration, which can be regarded as a specific form of international and intergovernmental cooperation, has triggered manifold changes in the Baltic Sea Region. As already mentioned earlier, the enlargement of the European Union had an impressive impact upon international environmental governance in the region. The 1995 enlargement of the EU to include the Nordic countries of Sweden and Finland strengthened the influence of the Nordic countries in the European Union. The EU enlargement in 2004 was probably even more important for the international environmental governance of the Baltic Sea Region, mainly for the following three reasons. First, the Baltic Sea became a domestic sea of the European Union. Only small – although important – parts now remain outside the sphere of the EU: the Russian Oblast Leningrad, with the metropolitan city of St Petersburg, and the Russian enclave Kaliningrad. The EU membership of Poland, Lithuania, Latvia and Estonia has facilitated coordination in the region considerably.

Second, the Europeanization of the region has been caused by several factors; but increasing economic interaction between the former communist countries and the West has played an important role in this respect. As diffusion studies show, increasing economic interaction leads to an adjustment of policies (Simmons and Elkins, 2004; Jahn, 2006). This is true for many policies in the countries in Central and Eastern Europe (Andonova, 2004; Jahn and Müller-Rommel, 2008). The trade exchanges of the four former communist EU member states in the Baltic Sea Region show quite clearly that they are oriented towards Western countries. Trade with Russia has declined in all former communist countries. Germany has become an important trading partner for all countries and, in particular, for Poland. Estonia is an exception in this respect as the Nordic countries, especially Finland, are its most important trading partners.

Third, the most important driving force for the Europeanization of the region has been EU legislation. The EU influences its member states in various ways (Knill and Lehmkuhl, 1999), and compliance with its legislation differs considerably between them. Although Europeanization can be regarded as co-evolution between the domestic and European level (Radaelli, 2006, p59), enlargement puts countries that wish to join the union under extreme pressure. The EU strongly influenced the new member states of Poland, Latvia, Lithuania and Estonia (and, of course, the other new member states in Central and Eastern Europe) in the pre-accession stage and put more pressure on them to adjust to European standards than in any other accession process. Before joining the EU, the countries of Central and Eastern Europe had to comply with the *acquis communautaire*, the entire body of EU legislation. This adjustment pressure has been referred to as 'forced diffusion', 'governance by conditionality' and 'diffusion by coercion' (Schimmelfennig and Sedelmeier, 2004; Braun and Gilardi, 2006). It would appear that EU influence on the countries of Central and Eastern Europe and the degree of legislative compliance was relatively high in the area of environmental policy (Andonova, 2005).

Transnationalization

Apart from Europeanization, the most striking characteristic of the Baltic Sea Region is its high degree of transnationalization. Since the end of the Cold War, the Baltic Sea Region has developed into a highly dynamic area of cross-border cooperation and transnational networking in the field of environmental protection. Not only have the traditional forms of governing focused on this policy area, but various new organizational structures, that can best be described as hybrid forms of governance (including governmental, sub-national and non-governmental actors), are also emerging at a rapid pace in the Baltic Sea Region (Joas et al, 2007, p241).

Three forms of transnationalization exist in the Baltic Sea Region. First, traditional international and intergovernmental organizations such as HELCOM have been transformed during recent years. Access to decision-making has improved for non-governmental and sub-national actors, and a more open governance structure has developed. Second, new types of organizations were established that aimed from the outset at bringing non-governmental and sub-national actors into the policy-making process. An outstanding example is Baltic 21, a regional Agenda 21 process for the whole Baltic Sea Region. Third, transnational networks, such as the Union of the Baltic Cities (UBC), which has more than 100 members in all countries around the Baltic Sea, work intensively with sustainability and environmental issues. The main goals of such networks are best practice transfer and learning among their members, and representation and lobbying. Networks such as the UBC run projects that establish direct links between their members and facilitate both transboundary policy transfer and the joint development of innovative solutions. The UBC's Agenda 21 Action Programme certainly helped to spread LA21 processes in the region. Transnational networks are also created to bypass nation states and are well prepared to establish direct contacts between the network and the European Union (especially the European Commission) (see Chapter 6 in this volume).

Governance for sustainable development in the Baltic Sea Region undoubtedly requires a combination of national governance with governance beyond the nation state. In this respect, transnational networks with contacts to the European Union provide promising new approaches that can complement the traditional forms of international and intergovernmental cooperation between nation states.

COMPARING THE BALTIC SEA WITH OTHER REGIONAL SEAS

Characteristics of comparable regions

The Baltic Sea shares certain features with other regional seas, such as the Mediterranean Sea and the Black Sea, which are both influenced by European integration and are, therefore, of interest from a comparative perspective. Like the Baltic Sea, the Black Sea has been directly affected by the end of the Cold War. Many of the riparian states, including Russia, are in transition from socialist states to market-oriented economies. Furthermore, the Baltic Sea can be compared with many other regional seas around the world. Since we cannot provide a comprehensive overview here, we will focus on two examples: the Great Lakes in North America and the Great Lakes in Africa. All of these regional seas are totally or almost entirely isolated from oceans and show slow rates of replenishment and long retention times for toxins. Such seas are extremely sensitive to accumulated contaminants and pollutants. Most of these regional seas are also heavily populated, thus increasing environmental pressures on ecosystems from both industrial and agricultural production. Therefore, environmental problems have become evident in all regions and have triggered the establishment of new institutional arrangements and the development of new policy approaches (for the Baltic Sea Region, see Chapter 9).

In all of these regional seas, multilevel structures have been established. Policy goals that are often formulated at the international level have direct impacts upon local activities (for the Baltic Sea Region, see Chapter 8 in this volume). European integration, which goes even one step further, has become a strong force in the Mediterranean and in the Black Sea Region (for similar developments in the Baltic Sea Region, see Chapter 6 in this volume; Kern and Löffelsend, 2004). Moreover, trends towards transnationalization, which are apparent in the Baltic Sea Region, can also be observed in other regional sea areas.

International and intergovernmental cooperation

Although the Mediterranean Sea Region has always been characterized by differing economic, social and environmental conditions at its northern and southern shores, initial efforts to improve the environmental situation started relatively early. These efforts were triggered by the creation of the Regional Seas Programme, an initiative which was started by the United Nations Environment Programme (UNEP) in 1974. Today this programme covers 18 regions in the world, including the Black Sea Region. The riparian states of the Mediterranean Sea and the EC/EU adopted the Mediterranean Action Plan in 1975 and the Convention for the Protection of the Mediterranean Sea against Pollution (Barcelona Convention) one year later (for details, see Haas, 1990). In 1996, the Mediterranean Commission on Sustainable Development was established. Supported by the Euro–Mediterranean partnership and several UN agencies, the Mediterranean Strategy for Sustainable Development (Agenda MED 21) was finally adopted in 2005. The preparation of this strategy stimulated the development of many national sustainable development strategies (e.g. in Morocco, Egypt and Syria) (Hoballah, 2006, pp157–164). Recently the European Investment Bank, in cooperation with the EU, UNEP and the World Bank, launched a Mediterranean Hot Spot Investment Programme, which aims to reduce pollution in the Mediterranean Sea by identifying and tackling the most significant pollution sources by the year 2020.

In the Black Sea, such initiatives were started much later. Although this regional sea has experienced a deep environmental crisis since the 1970s and is still one of the most polluted and environmentally degraded regional seas worldwide, a binding treaty for the protection of the Black Sea – the Bucharest Convention on the Protection of the Black Sea against Pollution – was adopted only in 1992. Although adopted in the same year as the Rio Declaration and Agenda 21, no reference to the principle of sustainable development can be found in the convention due to the fact that the text was created years before when the political climate did not favour its adoption (Doussis, 2006, p361). In the same year, the Commonwealth of Independent States (CIS) signed an inter-republican agreement on cooperation in the area of ecology. The Black Sea Environmental Programme (BSEP) was launched in 1993. It provided funding by the Global Environment Facility (GEF) and international donors. These efforts finally led to the Black Sea Strategic Action Plan for the Rehabilitation and Protection of the Black Sea, which was signed by the Black Sea countries in 1996 and revised in 2002. This plan defines the policy measures, actions and timetables required to achieve the environmental objectives of the Bucharest Convention. Many problems remain in the Black Sea Region due to limited cooperation among coastal states, the lack of funds and the difficulties in reconciling the interests of stakeholders who utilize the sea (Doussis, 2006).

In the Great Lakes Region in North America, the largest freshwater system on Earth, cooperation began much earlier than in the Baltic Sea Region, the Mediterranean Region or the Black Sea Region. The Boundary Waters Treaty of 1909 became the first agreement between the US and Canada. It also established the International Joint Commission (IJC), which held the authority to resolve disputes over the use of water resources that cross the boundary between the US and Canada. As early as 1919, the IJC detected serious water quality problems in this region and came to the conclusion that controlling pollution would require a new treaty. However, no agreement was reached until 1972 when the Great Lakes Water Quality Agreement was finally signed. The agreement was renewed in 1978 and amended in 1987. These revisions strengthened the ecosystem management approach of the agreement and provided the basis to review remedial action plans and lake-wide management plans. Because of their obligations under the agreement, both governments have established special programmes for the Great Lakes. The state of the environment in the region has improved considerably since the 1970s and the ecosystem of the Great Lakes has shown signs of recovery; but environmental pollution will certainly remain a major concern in the future (Klinke, 2006; Sage and Lynch, undated).

The Great Lakes in North America share some characteristics with the Great Lakes in East Africa. Lake Victoria is the second largest inland lake in the world, with an area close to 69,000 square kilometres. It is the most important freshwater reserve in the area, but rather shallow, thus highly vulnerable to pollution and ecological threats (for details, see www.gefweb.org). The shoreline of the lake is heavily populated, with approximately 30 million inhabitants in the basin area (see www.eac.int/lvdp/). The most important environmental threats to Lake Victoria are rapid population growth and extensive eutrophication with consequent algae blooms and loss of biodiversity due, for example, to extensive fishing (see www. gefweb.org). The largest share of Lake Victoria's shoreline belongs to Uganda and the largest surface area to Tanzania. The third country with a direct shoreline to the lake is Kenya. Additionally, Rwanda and Burundi constitute some 20 per cent of the basin area (see Klohn and Andjelic, undated).

The fact that Lake Victoria is one of the most important sources of Nile River waters and the fact that the region belonged to the British Empire before independence has resulted in some international regulations of the usage of natural resources in the area at an early stage. Several attempts to regulate and monitor the fisheries in the lake have also been established during the last 50 years. Riparian countries have all acknowledged the importance of environmental protection of Lake Victoria in their national environmental protection plans, and the three shoreline countries have been cooperating in this protection effort since 1992. This activity is coordinated with an international tripartite agreement, the Lake Victoria Environmental Management Programme, signed in August 1994 (see www.gefweb.org). This institution is today embedded in the framework of the East African Community, since 1999 a regional intergovernmental organization of Kenya, Uganda and Tanzania, with headquarters in Arusha, Tanzania (see www. eac.int/).

For all regional seas discussed here, international conventions and treaties have been developed and implemented. However, marked differences between the Helsinki Convention and its counterparts in other regional sea areas exist. These conventions were adopted at different points in time. Apart from early development in the North American Great Lakes Region, the Helsinki Convention was signed earlier (1974) than other comparable conventions and became the first international regime on the protection of a regional sea due to both increasing environmental problems and political reasons (see Chapter 3 in this volume). While the Barcelona Convention for the Protection of the Mediterranean Sea was already signed in 1976, the development of the Bucharest Convention for the Protection of the Black Sea and the Lake Victoria Environmental Management Programme followed only in 1992 and 1994.

These differences may be explained by structural factors. First, the number of riparian states varies considerably. It is lowest for the North American Great Lakes Region (US and Canada) and highest for the Mediterranean Sea Region (21 countries). Second, economic, political and cultural differences between the riparian states are most limited in the Great Lakes Region in North America and strongest for the Mediterranean Sea. While two highly industrialized and democratic countries share a lake system in North America, the Mediterranean countries are diverse and their economic development varies from highly industrialized countries to very poor countries.

It appears that the 1992 Rio Conference had a much stronger impact upon the Baltic Sea Region than upon other regional sea areas. This event stimulated not only many Agenda 21 processes at national and local levels (for the development of LA21 processes in the Baltic Sea Region, see Chapter 7 in this volume), but also the development of Baltic 21, the first regional Agenda 21 process worldwide. A comparable programme was launched only for the Mediterranean Sea, although only seven years later than in the Baltic Sea Region. This may be explained by the fact that the Nordic countries are environmental pioneers, while the former communist countries in the region were the most developed countries in their bloc.

EU integration and Europeanization

European integration has become very important for governance of the Mediterranean and the Black Sea Region because the recently launched European Neighbourhood Policy (ENP) applies to both regions. In the Mediterranean Region, EU integration and enlargement have improved the economic and political situation of Mediterranean EU member states (Greece, Spain, Slovenia, Malta and Cyprus). Furthermore, the Barcelona Process for the Mediterranean Region (Barcelona Declaration or Euro–Mediterranean Partnership), which began in 1995, became the EU's first regional initiative. It combines traditional security issues (proliferation of nuclear weapons, terrorism, etc.) with new security issues (energy security, environmental security, etc.). Since then the EU has become a driving force in the region. Furthermore, the EU developed a Marine Strategy and the European Commission proposed a Marine Strategy Directive in 2005. The EU Marine Strategy will require EU member states in all regional seas bordered by the EU to ensure cooperation with all countries in the region. To this end, member states will be encouraged to work within the framework of regional seas conventions.

In the Black Sea Region, the EU's involvement started much later and can be regarded as a direct consequence of the EU's enlargement in January 2007, when Bulgaria and Romania joined the union. The Black Sea Region has gained in strategic importance for the European Union because it has become the main corridor between Europe, Central Asia and the Middle East. In socio-economic, cultural and environmental terms, the Black Sea Region is one of the most sensitive areas in Europe (Güneş-Ayata et al, 2005). Furthermore, the European Commission launched the Black Sea Synergy in April 2007, which aims at a better economic, political and environmental cooperation between the states in the region. Black Sea Synergy is also a new regional cooperation initiative of the EU. Like the Euro–Mediterranean partnership and the Northern Dimension, its focus is better cooperation between the EU and its direct neighbours. Sectors mentioned in the Black Sea Synergy include, *inter alia*, energy, environment, transportation, security and democratic governance. Environment appears to be the only sector where successful cooperation projects already exist. The EU sees a need to enhance implementation of multilateral environmental agreements and to establish more strategic cooperation in the region. Moreover, the EU Commission declared community accession to the Bucharest Convention a priority. The European Commission already chairs the Danube Black Sea Task Force, which was set up by the countries of the Danube–Black Sea Region in 2001 to encourage a strategic focus on investment in the field of water management.

If the Baltic Sea Region is compared with both the Mediterranean Sea and the Black Sea Region, important differences become transparent. While the Baltic Sea Region changed tremendously after the EU's enlargement in 1995 and 2004, similar developments in the Mediterranean and the Black Sea Region cannot be expected. The only countries which have gained the status of candidate countries and may eventually join the EU in the future are Turkey and the East Adriatic countries. In contrast to the Baltic Sea, which is – with Russia as the only exception – surrounded almost only by EU member states, most of the riparian states of the Mediterranean and the Black Sea (such as Morocco or Georgia) will likely never become full EU member states. For both regions, the European Neighbourhood Policy will determine future developments. The ENP seeks to deepen economic integration and political cooperation between the EU and its immediate neighbours, which may in the long run foster sustainable development in both the Mediterranean and the Black Sea Region.

Transnationalization: Sub-national governments and civil society

Apart from international agreements and Europeanization, the 'relative success' of governance for sustainable development in the Baltic Sea Region can be explained by transnational factors. This includes both sub-national governments (regions and local authorities) and civil society actors. The Baltic Sea Region appears to be a fertile ground for the emergence of these different types of transnational networks.

Transnational municipal networks have developed and thrived in Europe since the early 1990s (Kern and Bulkeley, 2008). In contrast to other similar networks, the Union of the Baltic Cities was not launched by the EU or national governments. Instead, it developed spontaneously soon after the fall of the Berlin Wall. This network has grown considerably in the early years of its existence and has long reached a stabilization phase. Similar developments have not occurred in the Mediterranean and in the Black Sea Region, where city cooperation is more limited and in the early stages. Networks such as Medcities, a transnational city network in the Mediterranean Sea Region, have fewer members (only one or two per country in the case of Medcities) and are far less organizationally developed than the Union of the Baltic Cities. This may change in the future because the

Council of Europe's Congress of Local and Regional Authorities has launched a new initiative which aims at an increase in cooperation between local and regional authorities. The congress pursues a strategy of launching a new generation of Euroregions based on the EU's new structural policy, which includes European territorial cooperation as one of its three objectives.

Cooperation between local authorities can also be found in the Lake Victoria region where environmental cooperation has, supported by the Swedish government, among others, moved forward by introducing extensive local government-level networking. The Lake Victoria Region Local Authorities Cooperation (LVRLAC) was initiated as early as 1997, and it is essentially a regional network of local authorities around Lake Victoria. The main scope of the network is to work within and together with local communities at the lake to ensure sustainable use of resources. Membership counts up to 62 local governments in all three shoreline countries – Tanzania, Kenya and Uganda (see www.lvrlac.net). In addition, the network works with associated partners in order to gain both funding and knowledge. The Swedish Development Agency, together with the Union of the Baltic Cities, has been involved in a knowledge diffusion project, based on the experience and the good results gained in the Baltic Sea Area.

Focusing on the transnationalization of civil society, similarities between the regions can also be found. It appears that civil society has developed into transnational NGOs who support sustainable development in all regional sea areas discussed here. An example is the Mediterranean programme of the World Conservation Union (IUCN), established in 1996. IUCN has more than 170 members in the Mediterranean Region, including not only most states bordering the Mediterranean Sea, but also national and international NGOs. In the Black Sea Region, the Black Sea NGO Network (BSNN), a regional association of 62 NGOs from all Black Sea countries, was founded in 1998 and financially supported by the Global Environment Facility. Moreover, the European Commission has established a Black Sea Cross-Border Cooperation Programme under the European Neighbourhood Policy. This 'sea basin programme' focuses on supporting civil society and local-level cooperation in the Black Sea Region. The programme will be managed locally in the region, with partners taking joint responsibility for its implementation. Most environmental NGOs in the Black Sea Region appear to be externally funded – namely, by the EU and other international organizations (Aybak, 2005, p34).

The early development of Baltic 21 appears to be another rather unique feature of the Baltic Sea Region. Similar institutional arrangements have not been established in all other regional sea areas discussed here. Public dialogues with both the general public and stakeholder groups can only be found in the North American Great Lakes Region. The International Joint Commission holds public meetings every two years to discuss progress in cleaning up the Great Lakes. The stakeholder group consists of representatives from all levels of government, industry, environmental groups and academia (Sage and Lynch, undated). In the Mediterranean Sea Region, the Mediterranean Commission on Sustainable Development (MCSD) was set up and a Sustainable Development Strategy was adopted in 2005. As in the Baltic Sea Region, the preparation of this strategy was based on a participatory and multi-stakeholder approach and included the active participation of civil society (Hoballah, 2006, p163).

CONCLUSIONS

If the Baltic Sea Region is compared with other regional seas, we find many similarities but also marked differences. Because of European integration processes, the Baltic Sea Region can best be compared to the Mediterranean and the Black Sea Region. The Great Lakes Region in North America is difficult to compare because it comprises only two nation states: the US and Canada. This institutional setting facilitates cooperation in this region considerably.

In contrast to the Baltic Sea Region, for both the Mediterranean and the Black Sea Region, EU enlargement will have only limited effects. For both regions, the European Neighbourhood Policy will be more important instead. Europeanization in these regions is primarily influenced by this new EU approach, developed especially for neighbouring countries which have not yet (and will most likely never) become candidate countries and eventually full EU member states.

In the Baltic Sea Region, institutional arrangements exist which appear to be rather unique features of this region. Only in the Mediterranean Sea Region has a counterpart to Baltic 21 been established, although this happened much later than in the Baltic Sea Region. Furthermore, transnational city networking, which has a long tradition in the Baltic Sea Region, appears to be further developed than in the other regions discussed here. The early development of transnational city networking in the Baltic Sea Region can, at least partially, be explained by the historic development of the region.

Comparison with other similar regions shows that the Baltic Sea Region is special because a common sea is managed by a limited number of highly advanced industrial societies. How they do it can make us optimistic. After European unification, the Baltic Sea Region became a model for European integration. New forms of governance and institutional arrangements, which were developed in the Baltic Sea Region since the end of the Cold War, may now be transferred to comparable regions and may subsequently open new chances for sustainable development in these regions.

OUTLOOK

In social sciences, new insights of understanding seldom appear quickly. The development of science is based on small-scale developments: small steps in one direction, and perhaps somewhat larger steps in another direction. The build-up of new knowledge is by definition based on the understanding of these smaller and larger steps – thus stepwise in itself.

In this project we have been following an essentially small-scale societal development evident in most parts of the industrialized and democratic world, and obviously very visible in the Baltic Sea Region. This development is, however, highly needed in the Baltic Sea Region, as well as in other vulnerable and threatened marine and freshwater ecosystems.

In this chapter we have compared regional sea areas and highlighted how traditional governance structures changed into new forms of governance due to environmental pressure, often involving more stakeholders and more citizens. Cross-border cooperation at all governmental and societal levels can give us an opportunity to improve existing institutional arrangements and save some of the most vulnerable marine systems in the world.

REFERENCES

Andonova, L. (2004) *Transnational Politics of the Environment: The European Union and Environmental Policy in Central Eastern Europe*, MIT Press, Cambridge, MA

Andonova, L. (2005) 'The Europeanization of environmental policy in Central Eastern Europe', in F. Schimmelfennig and U. Sedelmeier (eds) *The Europeanization of Central and Eastern Europe*, Cornell University Press, Ithaca, pp135–155

Aybak, T. (2005) 'Interregional cooperation between the EU and BSEC', in A. Güneş-Ayata, A. Ergun and I. Çelimli (eds) *Black Sea Politics: Political Culture and Civil Society in an Unstable Region*, Tauris, London and New York, pp24–38

Braun, D. and Gilardi, F. (2006) 'Taking "Galton's problem" seriously: Towards a theory of policy diffusion', *Journal of Theoretical Politics*, vol 18, no 3, pp298–322

Doussis, E. (2006) 'Environmental protection of the Black Sea: A legal perspective', *South East European and Black Sea Studies*, vol 6, no 3, pp355–369

Güneş-Ayata, A., Ergun, A. and Işil, Ç. (eds) (2005) *Black Sea Politics: Political Culture and Civil Society in an Unstable Region*, I. B. Tauris, London and New York

Haas, P. (1990) *Saving the Mediterranean: The Politics of International Environmental Cooperation*, Columbia University Press, New York

Hermanson, A.-S. and Jahn, D. (1998) 'The state of the environment in West Europe', in J. Jabbra and O. Dwivedi (eds) *Governmental Response to Environmental Challenges in Global Perspective*, IOS Press, Amsterdam, pp95–109

Hoballah, A. (2006) 'Sustainable development in the Mediterranean Region', *Natural Resources Forum*, vol 30, pp157–167

Jahn, D. (2006) 'Globalization as Galton's problem: The missing link in the analysis of the diffusion patterns in welfare state development', *International Organization*, vol 60, no 2, pp401–431

Jahn, D. and Müller-Rommel, F. (2008) *Parlamentarische Demokratien in Mittelosteuropa: Institutionalisierung, Kondolidierung und Leitsungsbilanz*, University of Greifswald and Lüneburg, DFG Abschlussbericht

Jänicke, M. and Weidner, H. (eds) (1997) *National Environmental Policies: A Comparative Study of Capacity Building*, Springer, Berlin

Joas, M., Kern, K. and Sandberg, S. (2007) 'Actors and arenas in hybrid networks: Implications for environmental policymaking in the Baltic Sea Region', *Ambio*, vol 36, no 2–3, pp237–242

Kern, K. and Bulkeley, H. (2008) 'Cities, Europeanization and multi-level governance: Governing climate change through transnational municipal networks', *Journal of Common Market Studies* (forthcoming)

Kern, K. and Löffelsend, T. (2004) 'Sustainable development in the Baltic Sea Region: Governance beyond the nation state', *Local Environment*, vol 9, no 5, pp451–467

Kern, K., Koll, C. and Schophaus, M. (2007) 'The diffusion of Local Agenda 21 in Germany: Comparing the German federal states', *Environmental Politics*, vol 16, no 4, pp604–624

Klinke, A. (2006) *Demokratisches Regieren jenseits des Staates. Deliberative Politik im nordamerikanischen Große Seen-Regime*, Verlag Barbara Budrich, Opladen

Klohn, W. and Andjelic, M. (undated) *Lake Victoria: A Case in International Cooperation*, www.fao.org/ag/AGL/AGLW/webpub/lakevic/LAKEVIC4.htm, accessed 17 August 2007

Knill, C. and Lehmkuhl, D. (1999) 'How Europe matters: Different mechanisms of Europeanization', *European Integration Online Paper*, vol 3, no 7, http://eiop.or.at/eiop/texte/1990–007a.htm

Meadowcroft, J. (2007) 'National sustainable development strategies: Features, challenges and reflexivity', *European Environment*, vol 17, pp152–163

Radaelli, C. (2006) 'Europeanization: Solution or problem?', in M. Cini and A. Bourne (eds) *Palgrave Advances in European Union Studies*, Palgrave Macmillan, Basingstoke and New York

Sage, S. and Lynch, S. (undated) *Lessons from International Management of the Great Lakes*, www.aslf.org/ASLF/Pubs/The_Black_Sea2.html, accessed 6 August 2007

Schimmelfennig, F. and Sedelmeier, U. (2004) 'Governance by conditionality: EU rule transfer to the candidate countries of Central and Eastern Europe', *Journal of European Public Policy*, vol 11, no 4, pp669–687

Simmons, B. and Elkins, Z. (2004), 'The globalization of liberalization: Policy diffusion in the international political economy', *American Political Science Review*, vol 98, no 1, pp171–189

Steuer, R. (2007) 'From government strategies to strategic public management: An exploratory outlook on the pursuit of cross-sectoral policy integration', *European Environment*, vol 17, pp201–214

Internet sites

East African Community: www.eac.int/, accessed 17 August 2007

Lake Victoria Development Programme: www.eac.int/lvdp, accessed 17 August 2007

Lake Victoria Environmental Management Project Proposal: www.gefweb.org/COUNCIL/council7/wp/lakevic.htm, accessed 17 August 2007

Lake Victoria Region Local Authorities Cooperation (LVRLAC): www.lvrlac.net/, accessed 17 August 2007

Index